Lab 206

KU-784-844

OLIGONUCLEOTIDES
AS THERAPEUTIC
AGENTS

The Ciba Foundation is an international scientific and educational charity (Registered Charity No. 313574). It was established in 1947 by the Swiss chemical and pharmaceutical company of CIBA Limited — now Ciba-Geigy Limited. The Foundation operates independently in London under English trust law.

The Ciba Foundation exists to promote international cooperation in biological, medical and chemical research. It organizes about eight international multidisciplinary symposia each year on topics that seem ready for discussion by a small group of research workers. The papers and discussions are published in the Ciba Foundation symposium series. The Foundation also holds many shorter meetings (not published), organized by the Foundation itself or by outside scientific organizations. The staff always welcome suggestions for future meetings.

The Foundation's house at 41 Portland Place, London W1N 4BN, provides facilities for meetings of all kinds. Its Media Resource Service supplies information to journalists on all scientific and technological topics. The library, open five days a week to any graduate in science or medicine, also provides information on scientific meetings throughout the world and answers general enquiries on biomedical and chemical subjects. Scientists from any part of the world may stay in the house during working visits to London.

http://www.cibafoundation.demon.co.uk

Ciba Foundation Symposium 209

OLIGONUCLEOTIDES AS THERAPEUTIC AGENTS

1997

JOHN WILEY & SONS

Chichester · New York · Weinheim · Brisbane · Toronto · Singapore

Copyright © Ciba Foundation 1997
Published in 1997 by John Wiley & Sons Ltd,
　　　　　　　　Baffins Lane, Chichester,
　　　　　　　　West Sussex PO19 1UD, England

　　　　　　　　National 01243 779777
　　　　　　　　International (+44) 1243 779777
　　　　　　　　e-mail (for orders and customer service enquiries): cs-books@wiley.co.uk
　　　　　　　　Visit our Home Page on http://www.wiley.co.uk
　　　　　　　　　　　or http://www.wiley.com

All Rights Reserved. No part of this book may be reproduced, stored in a retrieval
system, or transmitted, in any form or by any means, electronic, mechanical, photocopying,
recording or otherwise, except under the terms of the Copyright, Designs and Patents Act
1988 or under the terms of a licence issued by the Copyright Licensing Agency, 90
Tottenham Court Road, London, UK W1P 9HE, without the permission in writing of the
publisher.

Other Wiley Editorial Offices

John Wiley & Sons, Inc., 605 Third Avenue,
New York, NY 10158-0012, USA

WILEY-VCH Verlag GmbH, Pappelallee 3,
D-69469 Weinheim, Germany

Jacaranda Wiley Ltd, 33 Park Road, Milton,
Queensland 4064, Australia

John Wiley & Sons (Canada) Ltd, 22 Worcester Road,
Rexdale, Ontario M9W 1L1, Canada

John Wiley & Sons (Asia) Pte Ltd, 2 Clementi Loop #02-01,
Jin Xing Distripark, Singapore 129 809

Ciba Foundation Symposium 209
x + 250 pages, 63 figures, 8 tables

Library of Congress Cataloging-in-Publication Data

Symposium on Oligonucleotides as Therapeutic Agents (1997 : Ciba
　　Foundation)
　　　　Oligonucleotides as therapeutic agents / [editors, Derek J.
　　Chadwick and Gail Cardew].
　　　　　　p.　cm. — (Ciba Foundation symposium ; 209)
　　　　Proceedings of the Symposium on Oligonucleotides as Therapeutic
　　Agents, held at Ciba Foundation on 7–9 January 1997.
　　　　Includes index.
　　　　ISBN 0-471-97279-7 (alk. paper)
　　　　1. Oligonucleotides — Therapeutic use — Congresses.　2. Antisense
　　nucleic acids — Therapeutic use — Congresses.　I. Chadwick, Derek J.
　　II. Cardew, Gail.　III. Series.
　　RM666.N87S96　1997
　　615′.31—dc21　　　　　　　　　　　　　　　　　　　　　97–24989
　　　　　　　　　　　　　　　　　　　　　　　　　　　　　　　　CIP

British Library Cataloguing in Publication Data

A catalogue record for this book is available from the British Library

ISBN 0 471 97279 7

Typeset in 10/12pt Garamond by Dobbie Typesetting Limited, Tavistock, Devon.
Printed and bound in Great Britain by Biddles Ltd, Guildford and King's Lynn.
This book is printed on acid-free paper responsibly manufactured from sustainable forestation,
for which at least two trees are planted for each one used for paper production.

Contents

v

Participants

S. A. Agrawal   Hybridon Inc., 620 Memorial Drive, Cambridge, MA 02139, USA

S. Akhtar *(Bursar)*   Pharmaceutical and Biological Sciences, Aston University, Aston Triangle, Birmingham B4 7ET, UK

B. Calabretta   Department of Microbiology and Immunology, Thomas Jefferson University, Philadelphia, PA 19107, USA

M. H. Caruthers *(Chairman)*   Department of Chemistry and Biochemistry, University of Colorado, Boulder, CO 80309–0215, USA

J. Cohen   Sheba Medical Center, Tel Hashomer, 52621, Israel

S. T. Crooke   Isis Pharmaceuticals Inc., Carlsbad Research Center, 2292 Faraday Avenue, Carlsbad, CA 92008, USA

F. Eckstein   Max-Planck Institute of Experimental Medicine, Hermann-Rein-Strasse 3, D-37075 Göttingen, Germany

K. L. Fearon   Chemical Process Research & Development, Lynx Therapeutics Inc., 3832 Bay Center Place, Hayward, CA 94545, USA

M. J. Gait   Medical Research Council, Laboratory of Molecular Biology, Hills Road, Cambridge CB2 2QH, UK

A. M. Gewirtz   Departments of Pathology and Laboratory Medicine and Internal Medicine, University of Pennsylvania School of Medicine, Philadelphia, PA 19104, USA

R. Haener   Ciba-Geigy Limited, CH-4002 Basle, Switzerland

C. Hélène   Laboratoire de Biophysique, Muséum National d'Histoire Naturelle, INSERM U 201, CNRS URA 481, 43 Rue Cuvier, 75231, Paris Cédex 05, France

M. Inouye   Robert Wood Johnson Medical School, Department of Biochemistry, 675 Hoes Lane, Piscataway, NJ 08854–5635, USA

P. Iversen   University of Nebraska, Department of Pharmacology, 600 S 42nd Street, Omaha, NE 68198–6260, USA

A. Krieg   University of Iowa College of Medicine, Department of Internal Medicine, 540 EMRB, Iowa City, IA 52242, USA

B. Lebleu   Molecular Genetics Institute, CNRS, 1919 Route de Mende, BP 5051, F-34033 Montpellier, Cédex 1, France

R. Letsinger   Northwestern University, Department of Chemistry, 2145 Sheridan Road, Evanston, IL 60208, USA

M. Matteucci   Gilead Sciences Inc., 353 Lakeside Drive, Foster City, CA 94404, USA

B. P. Monia   Isis Pharmaceuticals Inc., Department of Molecular Pharmacology, 2292 Faraday Avenue, Carlsbad, CA 92008, USA

P. Nicklin   Ciba Pharmaceuticals, Wimblehurst Road, Horsham, West Sussex RH12 4AB, UK

E. Ohtsuka   Hokkaido University, Faculty Pharmaceutical Sciences, Sapporo, Hokkaido 060, Japan

W. Pieken   NeXstar Pharmaceuticals Inc., 2860 Wilderness Place, Boulder, CO 80301, USA

J. J. Rossi   Department of Molecular Biology, Beckman Research Institute of the City of Hope, 1450 E Duarte Road, Duarte, CA 91010-3011, USA

E. M. Southern   Department of Biochemistry, University of Oxford, South Parks Road, Oxford OX1 3QU, UK

C. A. Stein   Department of Medicine, Columbia University, College of Physicians and Surgeons, 630 W 168 Street, Black Building 20–07, New York, NY 10032, USA

J.-J. Toulmé   Université Victor Segalen, IFR Pathologies Infectieuses, INSERM U 386, 146 Rue Léo Saignat Bat 3A, F-33076 Bordeaux, France

V. Vlassov   Russian Academy of Sciences, Institute of Bioorganic Chemistry, RO-630090, Novosibirsk, Russia

R.W.Wagner   Gilead Sciences Inc., 353 Lakeside Drive, Foster City, CA 94404, USA

E.Wickstrom   Department of Microbiology and Immunology and Kimmel Cancer Centre, Thomas Jefferson University, 1025 Walnut Street, Philadelphia, PA 19107, USA

Introduction

Marvin H. Caruthers

Department of Chemistry and Biochemistry, University of Colorado, Boulder, CO 80309-0215, USA

This symposium explores the use of antisense and catalytic nucleic acids for regulating gene expression. Antisense nucleic acids are single-stranded RNAs or DNAs that are complementary to the sequence of their target genes. Catalytic nucleic acids or ribozymes are also complementary to their target but, in addition, contain an enzyme-like catalytic activity capable of cleaving the target. This use of complementary or antisense nucleic acids to regulate gene activity has developed from nothing to a major research activity in the space of a little less than 20 years, with most of the effort coming in the past decade. For example over 100 research papers were published last year using antisense DNA oligomers to down-regulate target mRNAs and study the biology involved in turning off different genes. Examples include studies ranging from inducing exon skipping by targeting antisense to exon splice recognition sequences in the dystrophin gene (Promano et al 1996), inhibiting the multidrug resistance gene in tumour cells (Hughes et al 1996), down-regulation of the oncogene *erbB2* (Vaughn et al 1996) and dissecting induction of gene expression of platelet-derived growth factor (PDGF) by inhibiting protein kinase C (PKC) ζ (Xu et al 1996). In addition to the vigorous research activity surrounding this new field, there are clinical studies underway that focus on the use of antisense oligonucleotides as therapeutic drugs. Several outlining the preliminary work leading to various drug candidates will be discussed at this symposium. These include the potential use of antisense DNA for refractory cytomegalovirus (CMV) retinitis in AIDS patients, as a treatment for Crohn's disease and human leukaemias, and as an anti-AIDS therapeutic.

The concept of using antisense oligonucleotides as a new class of therapeutic drugs is based upon our understanding of nucleic acid structure and function. It depends upon the ability of these oligomers to target DNA or RNA through Watson–Crick, Hoogsteen or reverse Hoogsteen base pairing. In the former case, the target is single-stranded DNA or RNA, whereas for the latter, it is the major groove of double-stranded DNA. Thus, the complementary or antisense base sequence interacts by hydrogen bonding with a segment of cellular RNA or DNA to inhibit gene expression. The early demonstration that nucleic acid hybridization is feasible (Gillespie & Spiegelman 1965) provided one of the most basic elements of the foundation supporting the antisense concept. There have been other enabling technologies as well. These include the ability to identify specific genes and gene products through recombinant technologies, cloning, manipulation of DNA with restriction enzymes, PCR procedures, DNA sequencing and the development of rapid efficient DNA synthesis methodologies. The first clear elaboration of the antisense concept was the elegant work of Zamecnik & Stevenson (1978). They reported how a synthetic deoxyoligonucleotide that was complementary to a sequence in the Rous sarcoma virus genome could be used as an antiviral reagent. They postulated that this oligomer inhibited replication by binding to viral RNA—a mechanism that has since been used to define the antisense approach. During the same period, the pioneering work of Miller and T'so also focused on developing the rationale and experimental approaches needed for defining the antisense approach (Barrett et al 1974, T'so et al 1983). For example, they showed that certain regions of RNA, specifically the translation initiation domain of mRNAs, were more susceptible to inhibition by antisense oligomers than many other sites on this molecule. They were also the first to demonstrate the advantages of using DNA analogues, specifically phosphate-modified derivatives, to solve problems such as nuclease susceptibility inherent in natural DNA. Following these early observations, much of the ensuing work with oligonucleotides established the major tenets outlining the approach and demonstrated the effect in model systems (Crooke 1996).

As a result of this research, several principles relating to the successful use of antisense DNA have been established, whereas others remain elusive. For example, the uptake of oligonucleotides by cells, tissues and animals continues to be a challenging research area. As will become evident from many of the studies we will discuss, simple delivery of oligomers to humans or animals intravenously or intraperitoneally is sufficient to observe an effect. However, targeted delivery to certain tissues by tagging oligomers might reduce the dosage and enhance viability. This possibility remains to be explored and will be discussed by several at this symposium. In contrast, delivery to cells in culture is an entirely different matter. Simply incubating antisense oligomers with cells may lead to an antisense effect but, in most cases, the DNA fails to enter cells or is observed only in punctated vesicles from which it is expelled into the external culture media. As we will learn here, the only reliable methods currently available involve microinjection directly into the cytoplasm or the use of cationic liposomes, which are toxic to many cell types and require serum-free media during uptake. There are, however, several approaches being investigated for solving this problem and some will be presented at this symposium. The importance of this work is worth emphasizing because many basic questions in cell biology could be addressed via the antisense approach if DNA could be delivered reliably and under controlled conditions to cells in culture. Until this is possible, the use of the antisense approach in basic research will remain problematic at best because of uncertainty surrounding delivery.

There are many other biochemical and biological challenges that must be overcome before antisense oligonucleotides can become viable research tools and therapeutic drugs. One is to understand the mechanisms whereby oligonucleotides modify gene function. Classically, synthetic DNA has been used to activate RNase H, which in turn degrades RNA complementary to the antisense oligomer at the site of hybridization — hence the name RNase H, a ribonuclease that recognizes the hybrid structure and degrades the RNA component. This still remains the only recognized mechanism whereby antisense oligomers are functional *in vivo*. Other approaches shown to work *in vitro*, such as hybridization arrest and inhibition via triplex formation, have yet to be proven viable *in vivo*. However, DNA has now been shown to exhibit non-antisense effects that closely mimic expected antisense results. As we will learn at this symposium, certain oligomers containing guanosine-rich sequences have antiproliferative properties and others having CG sequence elements activate a co-ordinate set of immune responses including both humoral and cellular immunity. These observations are discussed throughout this symposium and indicate how much still remains before we understand the diverse mechanisms whereby DNA interacts with cells, tissues and organisms. A related challenge is to identify criteria that confirm an antisense effect either in cell culture or animals. These will also be discussed, and they include investigations on measuring the specific expression of target proteins and mRNA, as well as demonstrating the lack of an effect with oligomers having one or more mismatched sequence elements.

Early in the development of this field it became obvious that natural DNA and RNA could not fulfil the requirements expected of an antisense molecule. These included stability against endogenous nucleases, targeted uptake by cells and tissues followed by localization into a cellular compartment where an antisense effect could be carried out, and the formation of a highly specific complex with cellular DNA or RNA. In attempts to meet these criteria, several hundred analogues have been synthesized but, as we will learn here, most are of limited value. For example, with the exception of phosphorothioate and phosphorodithioate analogues, none of the derivatives investigated to date activate RNase H — currently an essential criterion for antisense activity. However, despite being RNase H inactive, some analogues will be discussed, including the 2′-O-methyl, 2′-O-methoxyethyl and methylphosphonate compounds, that have proven to be potentially valuable because they can be used to form chimeric molecules having desirable cellular uptake, enhanced nuclease stability and favourable pharmacological properties. As we will also discuss, an additional important challenge is the scale-up of oligonucleotide chemical synthesis. Initially considered an impossible barrier to the use of oligonucleotides as therapeutic drugs, this problem has been overcome as oligomers can now be produced in kilogram amounts cheaply and in high purity.

Perhaps the most significant challenge facing the antisense field today is whether it will work. Can antisense oligomers be used to help patients overcome a disease? This question can only be answered in the clinic using appropriate pharmacological, toxicological and pharmacokinetic parameters. Several groups will report their clinical results, which persuasively argue that indeed the technology should be vigorously pursued and may work as expected.

I have tried to highlight the topics that will be discussed at this symposium. Generally, this field has passed through several phases ranging from extreme optimism, primarily because of the simplistic nature of the concept, to considerable scepticism, when difficulties arose on many fronts, such as poor uptake of oligomers by cells, lack of biological activity by most oligomer analogues and difficulties encountered interpreting results. The latter problem exists primarily because oligomers cause many cellular effects, not through antisense mechanisms, but because these molecules are polyanionic or have sequence elements that stimulate non-antisense results. However, sorting out these problems, which were not initially fully appreciated, takes time, and, as we will see during the course of this symposium, many have been clarified and others solved. As a consequence, there is much optimism surrounding the field at the present time and I suspect we will all return to our laboratories with renewed vigour and enthusiasm for what lies ahead.

References

Barrett JC, Miller PS, Ts'o POP 1974 Inhibitory effect of complex formation with oligodeoxyribonueleotide ethyl phosphotriesters on transfer ribonucleic acid aminoacylation. Biochemisty 13:4897–4910

Crooke ST 1996 Progress in evaluation of the potential of antisense technology. Antisense Res Dev 4:145–146

Gillespie D, Spiegelman SA 1965 A quantitative assay for DNA–RNA hybrids with DNA immobilized on a membrane. J Mol Biol 12:829–842

Hughes JA, Aronsohn AI, Arcutskaya AV, Juliano RL 1996 Evaluation of adjuvants that enhance the effectiveness of antisense oligonucleotides. Pharm Res 13:404–410

Promano ZA, Takeshima Y, Alimsardjono H, Ishii A, Takeda S, Matsuo M 1996 Induction of exon skipping of the dystropin transcript in lymphoblastoid cells by transfecting an antisense oligonucleotide complementary to an exon recognition sequence. Biochem Biophys Res Commun 226:445–449

Ts'o POP, Miller PS, Greene JJ 1983 Nucleic acid analogues with targeted delivery as chemotherapeutic agents. In: Cheng YC, Goz B, Minkoff M (eds) Development of target-orientated anticancer drugs. Raven Press, New York, p 189–206

Vaughn JP, Stekler J, Demirdji S et al 1996 Inhibition of the ERBB-2 tyrosine kinase receptor in breast cancer cells by phosphorothioate and phosphorodithioate antisense oligonucleotides. Nucleic Acids Res 24:4558–4564

Xu JH, Zutter MN, Santoro SA, Clark RAF 1996 PDGF Induction of alpha (2) integrin gene expression is mediated by protein kinase C-zeta. J Cell Biol 134:1301–1311

Zamecnik PC, Stevenson ML 1978 Inhibition of Rous sarcoma viral replication and cell transformation by a specific oligodeoxynucleotide. Proc Natl Acad Sci USA 75:280–284

Oligoncleotide analogues: an overview

Mark Matteucci

Gilead Sciences Inc., 353 Lakeside Drive, Foster City, CA 94404, USA

Abstract. The medicinal chemistry effort directed toward improving antisense and antigene oligonucleotides has synthesized a large number of phosphate, ribose and heterocyclic base analogues. The phosphorothioate linkage is currently still the analogue linkage of choice for antisense studies. This is despite many years of effort to find alternatives that overcome the limitations of phosphorothioates. A number of modifications to phosphate and ribose have resulted in enhanced binding to RNA as measured by T_m, but generally the biological effects have been less dramatic. Currently, the most potent oligonucleotides have capitalized on recruiting the cellular enzyme RNase H to perform sequence-specific destruction of a targeted RNA. Virtually all modifications to phosphate or ribose other than phosphorothioate result in the loss of this recruitment. Chimeric strategies have overcome this limitation. Affinity and potency can additionally be improved by modifying the Watson–Crick-pairing heterocycles. Recent years have brought much consensus in terms of the mechanism of action of antisense oligonucleotides. A controversial area is the ability of oligonucleotides to permeate cells in whole animals. This issue will determine if antisense and triple helix technology results in practical broad-based therapeutics.

1997 Oligonucleotides as therapeutic agents. Wiley, Chichester (Ciba Foundation Symposium 209) p 5–18

In recent years there has been an explosion of effort devoted to the synthesis of oligonucleotide analogues. Early attempts at sequence-selective gene inhibition with unmodified oligonucleotides were provocative and led to an analogue effort to improve properties such as nuclease stability, cellular permeation and potency. This chapter will not encompass the history of analogues, but rather focus on derivatives that currently appear to improve a property or solve a problem. As with most medicinal chemistry efforts, many analogues are synthesized; some become the focus of excitement for a time and few persist as truly valuable.

The oligonucleotide analogue arena started with replacements of the phosphodiester linkage. The replacement of the phosphodiester linkage 1a (Fig. 1) was identified early as a necessary requirement for advancing antisense oligonucleotides, since unmodified oligonucleotides are rapidly degraded by nucleases that recognize the phosphodiester linkage. The nuclease stability issue was

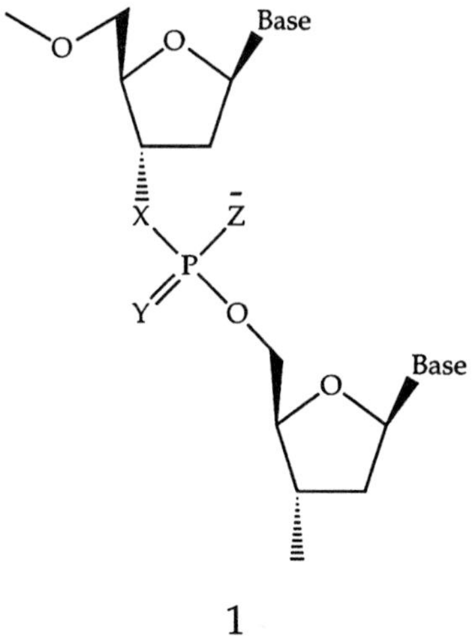

	X	Y	Z
a	O	O	O
b	O	O	S
c	O	S	S
d	O	O	BH$_3$
e	NH	O	O

1

FIG. 1.   Anionic phosphodiester analogues. For detailed explanation see text.

easily solved. Virtually any modification to the phosphate confers stability. Phosphate analogues can be divided into three groups based on their charge: anions, cations and neutrals. Anionic linkages bear the closest resemblance to the native phosphodiester linkage (Fig. 1).

The phosphorothioate linkage 1b has become the standard choice for most antisense attempts (Cohen 1993). The main advantages of phosphorothioates are stability to nucleases, ability to recruit RNase H and ease of synthesis. A disadvantage of the phosphorothioate linkage is the general lowering of affinity for the RNA target relative to a phosphodiester. This lower affinity is partially caused by the lack of stereocontrol during synthesis and the resulting mixture of diastereomers. In addition, this linkage appears to enhance oligonucleotide binding to protein. This phenomenon can be exploited to produce a receptor antagonist (Wyatt et al 1994) but the propensity of phosphorothioates to bind to protein is undesirable for antisense applications.

The phosphorodithioate linkage 1c was introduced to eliminate the chiral centre, thereby removing the diastereomer problem (Marshall & Caruthers 1993). The data available thus far do not demonstrate that this linkage has any advantage over the simple monothioate mentioned above. Its affinity for the target and ability to activate RNase H is similar to the diastereomeric mixtures of monothioates.

An intriguing analogue for which there are few binding data is the boranephosphite complex 1d (Sood et al 1991). This analogue is a surprisingly stable, chemically unique isostere of a phosphodiester. Recently, we synthesized a T$_{15}$ oligonucleotide bearing diastereomerically mixed boranephosphites at all positions and showed that it binds poorly to complementary RNA and DNA (L. Zhang & M. Matteucci, unpublished work 1996). This analogue, despite its elegant simplicity, is not likely to be useful.

Recently, the 3′ phosphoroamidate linkage 1e has attracted attention. Oligonucleotides bearing this achiral analogue at all positions are stable to nucleases *in vitro* and have a greatly elevated interaction with complementary RNA relative to phosphodiesters (Schultz & Gryaznov 1996). Preliminary cell culture data suggest that this analogue results in enhanced antisense potency (Gryaznov et al 1996). Additionally, this analogue is finding use in the triple helix arena because it greatly enhances third-strand binding to duplex DNA in the parallel pyrimidine–purine–pyrimidine motif (Escudé et al 1996).

All of these anionic linkages suffer from the problem of ineffective cellular permeation in culture. Anionic oligonucleotides traffic into the endosomes of cells in cell culture but they are ineffective at crossing the endosomal membrane in the absence of cationic lipids (Leonetti et al 1991). The simple intuitive notion that polyanions were undesirable for membrane permeability spurred efforts to identify neutral isosteres of phosphate. The first linkage to be extensively investigated was the methylphosphonate 2 linkage (Fig. 2; Miller 1991). The design goal behind the methylphosphonate analogue was that charge neutralization of the phosphate would lead to a more lipophilic molecule resulting in membrane permeability by passive diffusion — this is not the case. Methylphosphonate analogues suffer the same fate as phosphodiester and phosphorothioate analogues, and they are sequestered into vesicles (Shoji et al 1991).

The methylphosphonate analogue has the advantage of straightforward synthetic methods (Miller 1991). A deficiency of this linkage is its chiral nature. Different diastereomers have been shown to have significantly different binding affinities (Reynolds et al 1996). This observation has spurred attempts to obtain synthetic methodology that produces only one isomer. Diastereomerically pure linkages allow for improvement of binding affinity. It has not been determined whether or not diastereomerically pure molecules will result in improved biological potencies. The synthetic complexities of chiral synthesis, the lack of RNase H recruitment and questionable permeation cloud the value of the methylphosphonate analogue for the future.

The chiral nature of the methylphosphonate analogue and the desire to reduce or eliminate the polyanionic nature of oligonucleotides for permeation reasons has resulted in a plethora of achiral neutral substitutes for phosphate (De Mesmaeker et al 1995). There are a handful of linkages that appear to enhance or largely preserve affinity for a RNA target relative to phosphodiesters, these are shown in Fig. 2.

The most provocative members of the group are the peptide nucleic acid (PNA) (4, Fig. 2) and morpholinophosphorodiamidate (5, Fig. 2) linkages. These are of interest because of ease of synthesis and good binding properties. Formally, these derivatives go beyond a phosphate modification since the ribose moiety has been significantly altered. There has been an explosion of literature on the PNA front relating to their ability to bind to both RNA and duplex DNA via a strand-displacement mechanism

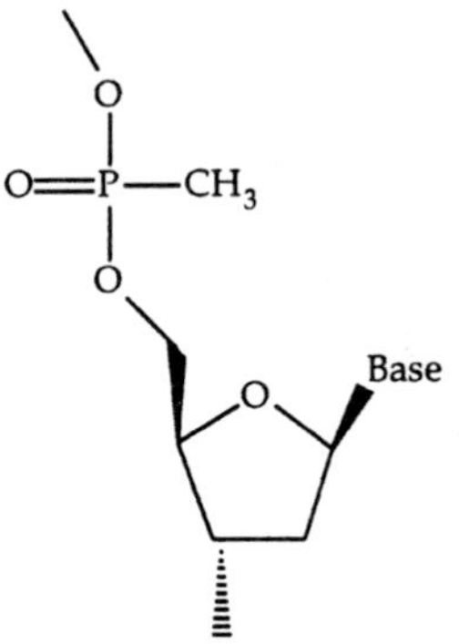

2

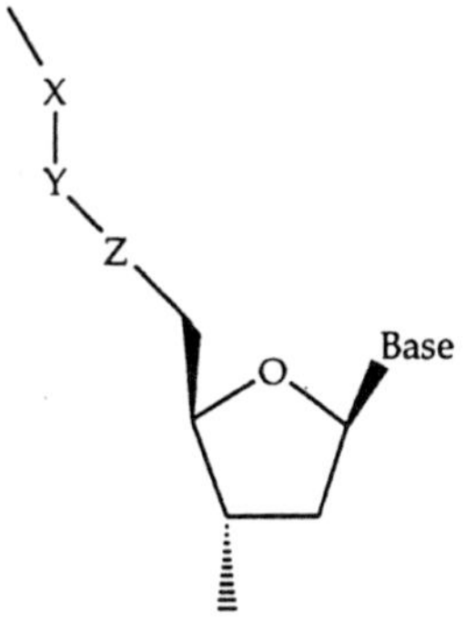

3

	X	Y	Z
a	S	CH_2	O
b	O	CH_2	O
c	CH_2	NMe	O
d	CH_2	C=O	NH
e	CH_2	SO_2	CH_2
f	NH	SO_2	CH_2
g	O	SO_2	NH

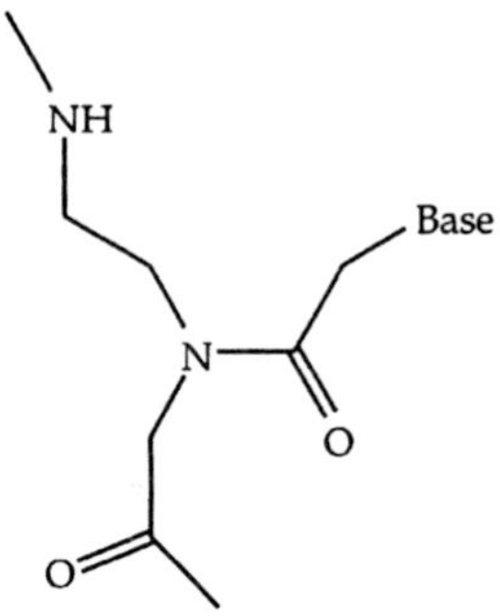

4

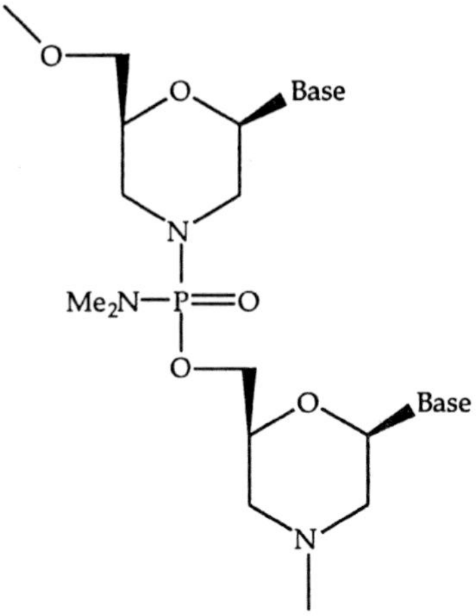

5

FIG. 2. Neutral backbones. For detailed explanation see text.

(Haaima et al 1996). What has been lacking in the PNA arena are data regarding the ability to regulate gene expression within a cell (Knudsen & Nielsen 1996, Bonham et al 1995). This is somewhat surprising in that one of the main virtues of PNA is the ease of synthesis, which allows biological questions to be answered quickly and without a synthetic *tour de force*.

The easily synthesized morpholinophosphoramidate modification is just beginning to receive attention (Summerton et al 1997). This derivative has been shown to possess

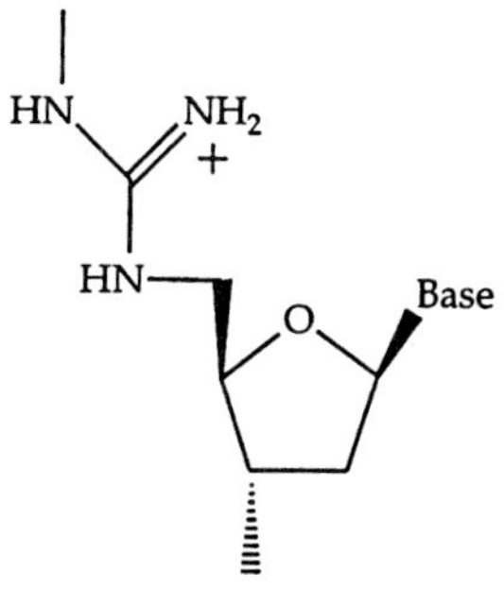

6

FIG. 3.   Cationic linkage. For detailed explanation see text.

intracellular antisense properties working by a non-RNase H mechanism. Again, as with all analogues to date, it suffers from the lack of effective cellular permeation in cell culture.

The last area of phosphate analogues is the cationic derivatives. The rationale for this series is to engineer an electrostatic attraction between the oligonucleotide and the polyanionic RNA target. In an ideal world this could enhance affinity without compromising specificity and improve cellular permeation. The achiral guanidine linkage (Fig. 3) has been shown in a pentathymine oligonucleotide to form extremely stable duplexes (Dempcy et al 1995). Further information regarding other contexts, cellular permeation and biological activity should be forthcoming.

The ribose portion of oligonucleotides has been extensively modified. This area has been reviewed, with a number of modifications showing provocative results (De Mesmaeker et al 1995). The ribose ring has been appended and substituted, conformationally restricted and even replaced by hexose. The most straightforward and perhaps most promising substitutions involve the modification of the 2' position (7, Fig. 4).

Analogue work in this area started with the 2'-O-methyl variant of RNA (Inoue et al 1987). This analogue was shown to confer enhanced binding to complementary sequences. The 2'-O-methyl group confers stability to single-stranded RNases but phosphodiester oligonucleotides are still susceptible to degradation by DNases. Highly electronegative substituents, such as fluorine, virtually lock the ribose ring in the RNA conformation leading to enhanced binding to RNA targets (Kawasaki et al 1993). However, such modifications do not stabilize phosphodiesters from nucleases and preclude RNase H cleavage of the RNA bound to the oligonucleotide because of their extreme RNA-like conformation.

Simple alkyl ethers have been synthesized to explore the effect of steric bulk at the 2' position. The trend is that nuclease stability increases and binding affinity decreases

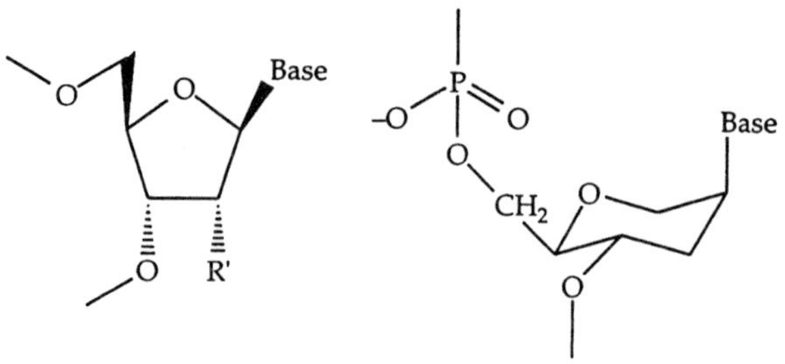

7   8

R = OMe, F, OCH₂CH₂OMe

FIG. 4.   Ribose analogues. For detailed explanation see text.

with increasing size (Monia et al 1996a). Recently, the introduction of the 2'-methoxyethoxy group has allowed one to have one's cake and eat it too (i.e. to have both nuclease stability with phosphodiesters and excellent binding affinity; Altmann et al 1996).

One provocative modification uses a hexose scaffold. The 1,5-anhydrohexitol oligonucleotide phosphodiesters (8, Fig. 4) hybridize well to RNA (Van Aerschot et al 1995). The altered sugar residue confers significantly enhanced affinities to RNA. This modification stabilizes a phosphodiester linkage toward nucleases. RNase H recruitment has not been addressed but given the extreme sensitivity of that enzyme to sugar pucker, the hexitol oligonucleotide–RNA hybrid will most likely not be a substrate for RNase H. The modified hexose is unlikely to improve the cellular permeation of diester oligonucleotides. Biological testing has not been reported at present.

The aforementioned properties of enhanced binding, nuclease stability and lack of RNase H recruitment are common to the ribose modifications described. These modifications, particularly the 2'-O-ether series, have been used in the gap or chimera strategy to overcome the lack of RNase H recruitment (Monia et al 1996a). These sequences contain a minimal core of about seven nucleotides of RNase H recruiting phosphorothioate and flanking sequences that are modified for nuclease stability and affinity. One recent example is the use of flanking regions of the 2'-methoxyethoxy modification in conjunction with phosphodiester linkages. This modification has resulted in enhanced antitumour activity when incorporated into a c-*raf* kinase oligonucleotide (Altmann et al 1996).

The aromatic heterocyclic bases are the recognition element of the oligonucleotide–RNA interaction. Extensive modifications to these bases have been made (Sanghvi 1993). A simple pyrimidine modification which is proving to be a general solution to enhanced affinity is the 5-propynyl uracil (9, Fig. 5) or cytosine (10, Fig. 5) derivative (Froehler et al 1992). These substitutions significantly enhance affinity.

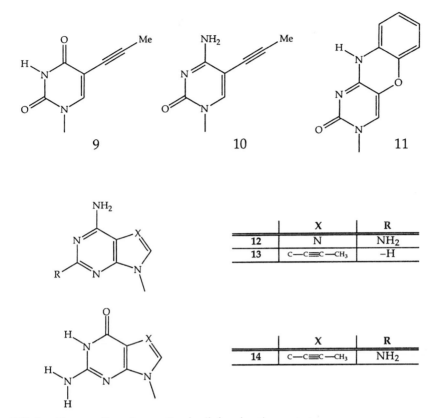

FIG. 5.   Heterocyclic analogues. For detailed explanation see text.

Antisense effects with propynyl pyrimidine oligonucleotides have been seen in a number of cell culture systems when the linkage is the nuclease stable phosphorothioate (Wagner et al 1993, 1996, Coats et al 1996, Flanagan et al 1996).

The most likely reason for the increased affinity of the propynyl series is stacking interactions between adjacent bases within a duplex and the lack of rotomers in the propyne group. Recent tricyclic analogues have attempted to maximize these pi overlaps. The phenoxazine analogue 11 (Fig. 5) results in stabilization of oligonucleotide–RNA duplexes (Lin et al 1995). The degree of stabilization is dependent on the context, with dramatic stabilization being observed when the tricyclic bases are localized in an adjacent fashion. This observation is consistent with the concept of aromatic ring overlap being the stabilizing force. Phenoxazine has conferred antisense potency comparable to 5-propynylcytosine substitution to some oligonucleotides in cell culture (R. W. Wagner & M. Matteucci, unpublished observations 1996).

Purines have been less investigated, largely because of their increased synthetic complexities. The simple purine analogue diaminopurine (12, Fig. 5) improves affinities relative to adenine (Gryaznov & Schultz 1994) due to an additional hydrogen bond being formed with the complementary uracil. Oligonucleotides containing 7-deazapurine analogues are known (Seela & Thomas 1995) and the 7-propynyl oligonucleotide derivatives of adenine (13, Fig. 5) and guanine (14, Fig. 5) have recently been shown to enhance binding to RNA (Buhr et al 1996). Curiously, only the guanine derivative improves the antisense activity in cell culture relative to the parent heterocycle. The class of highly lipophilic heterocyclic analogues has the potential of being more capable of permeating into the cytosol of cells. This is an area of continued study.

One of the most important and most controversial aspects of oligonucleotide analogues as it relates to the antisense and triple helix area is the issue of cellular permeation. Some will argue vehemently that permeation is not a problem, not in cell culture or in animals. Others argue that permeation is problematic only in cell culture with cationic lipids being a solution. They point to emerging animal studies and human trials as proof that *in vivo* permeation can occur using unassisted oligonucleotides. Others will argue the *in vivo* animal arena is less clear. They will argue that most animal studies have not been well controlled enough to allow for definitive conclusions. Observed phenotypes may be due to an antisense mechanism or some other effect elicited by a class of molecules known to have significant protein-binding properties.

Most investigators now agree that cellular permeation is usually extremely inefficient in cell culture without the use of a delivery technique such as cationic lipids (Bennett et al 1992, Lewis et al 1996). The consensus is that oligonucleotides, polyanions and neutrals alike are taken up into endosomes and are not released into the cytoplasm and nucleus of the cell. The controversy regarding permeation into cells within whole animals continues. It can only be settled with well-controlled data. Such studies are emerging (Monia et al 1996b) and will continue to emerge in the near future.

References

Altmann KH, Dean NM, Fabbro D et al 1996 Second generation of antisense oligonucleotides: from nuclease resistance to biological efficacy in animals. Chimia 50:168–176

Bennett CF, Chiang M-Y, Chan H, Shoemaker JE, Mirabelli CK 1992 Cationic lipids enhance cellular uptake and activity of phosphorothioate antisense oligonucleotide. Mol Pharmacol 41:1023–1033

Bonham MA, Brown S, Boyd AL et al 1995 An assessment of the antisense properties of RNase H-competent and steric-blocking oligomers. Nucleic Acids Res 23:1197–1203

Buhr CA, Wagner RW, Grant D, Froehler BC 1996 Oligodeoxynucleotides containing C-7 propyne analogs of 7-deaza-2'-deoxyguanosine and 7-deaza-2'-deoxyadenosine. Nucleic Acids Res 24:2974–2980

Coats S, Flanagan WM, Nourse J, Roberts JM 1996 Requirement of p27[kip1] for restriction point control of the fibroblast cell cycle. Science 272:877–880

Cohen JS 1993 Phosphorothioate oligodeoxynucleotides. In: Crooke ST, Lebleu B (eds) Antisense research, applications. CRC Press, Boca Raton, FL, p 205–221

De Mesmaeker AD, Häner R, Martin P, Moser HE 1995 Antisense oligonucleotides. Acc Chem Res 28:366–374

Dempcy RO, Browne KA, Bruice TC 1995 Synthesis of thymidyl pentamer of deoxyribonucleic guanidine and binding studies with DNA homopolynucleotides. Proc Natl Acad Sci USA 92:6097–6101

Escudé C, Giovannangeli C, Sun J-S et al 1996 Stable triple helices formed by oligonucleotide N3'→P5' phosphoramidates inhibit transcription elongation. Proc Natl Acad Sci USA 93:4365–4369

Flanagan WM, Su L, Wagner RW 1996 Elucidation of gene function using C-5 propyne antisense oligonucleotides. Nat Biotechnol 14:1139–1145

Froehler BC, Wadwani S, Terhorst TJ, Gerrard SR 1992 Oligodeoxynucleotides containing C-5 propyne analogs of 2'-deoxyuridine and 2'-deoxycytidine. Tetrahedron Lett 33:5307–5310

Gryaznov S, Schultz RG 1994 Stabilization of DNA:DNA and DNA:RNA duplexes by substitution of 2'-deoxyadenosine with 2'-deoxy-2'-aminoadenosine. Tetrahedron Lett 35:2489–2492

Gryaznov S, Skorski T, Cucco C et al 1996 Oligonucleotide N3'→P5' phosphoramidates as antisense agents. Nucleic Acids Res 24:1508–1514

Haaima G, Lohse A, Buchardt O, Nielsen PE 1996 Peptide nucleic acids (PNAs) containing thymine monomers derived from chiral amino acids: hybridization and solubility properties of D-lysine PNA. Angew Chem Int Ed Engl 35:1939–1942

Inoue H, Hayase Y, Imura A, Iwai S, Miura K, Ohtsuka E 1987 Synthesis and hybridization studies on two complementary nona(2'-O-methyl)ribonucleotides. Nucleic Acids Res 15:6131–6148

Kawasaki AM, Casper MD, Freier SM et al 1993 Uniformly modified 2'-deoxy-2'-fluoro phosphorothioate oligonucleotides as nuclease-resistant antisense compounds with high affinity and specificity for RNA targets. J Med Chem 36:831–841

Knudsen H, Nielsen PN 1996 Antisense properties of duplex- and triplex-forming PNAs. Nucleic Acids Res 24:494–500

Leonetti JP, Mechti N, Degols G, Gagnor C, Lebleu B 1991 Intracellular distribution of microinjected antisense oligonucleotides. Proc Natl Acad Sci USA 88:2702–2706

Lewis JG, Lin K-Y, Kothavale A et al 1996 A serum-resistant cytofectin for cellular delivery of antisense oligodeoxynucleotides and plasmid DNA. Proc Natl Acad Sci USA 93:3176–3181

Lin K-Y, Jones RJ, Matteucci M 1995 Tricyclic 2'-deoxycytidine analogs: syntheses and incorporation into oligodeoxynucleotides which have enhanced binding to complementary RNA. J Am Chem Soc 117:3873–3874

Marshall WS, Caruthers MH 1993 Phosphorodithioate DNA as a potential therapeutic drug. Science 259:1564–1570

Miller PS 1991 Oligonucleoside methylphosphonates as antisense reagents. Biotechnology 9:358–365

Monia BP, Johnston JF, Sasmor H, Cummins LL 1996a Nuclease resistance and antisense activity of modified oligonucleotides targeted to Ha-*ras*. J Biol Chem 271:14533–14540

Monia BP, Johnston JF, Geiger T, Muller M, Fabbro D 1996b Antitumor activity of a phosphorothioate antisense oligodeoxynucleotide targeted against C-*raf* kinase. Nat Med 2:668–675

Reynolds MA, Hogrefe RI, Jaeger JA 1996 Synthesis and thermodynamics of oligonucleotides containing chirally pure Rp methylphosphonate linkages. Nucleic Acids Res 24:4584–4591

Sanghvi YS 1993 Heterocyclic base modifications in nucleic acids and their applications in antisense oligonucleotides. In: Crooke ST, Lebleu B (eds) Antisense research, applications. CRC Press, Boca Raton, FL, p 273–288

Schultz RG, Gryaznov SM 1996 Oligo-2'-fluoro-2'-deoxynucleotide N3'→P5' phosphoramidates: synthesis and properties. Nucleic Acids Res 24:2966–2973

Seela F, Thomas H 1995 Duplex stabilization of DNA: oligonucleotides containing 7-substituted 7-deazaadenines. Helv Chim Acta 78:94–97

Shoji Y, Akhtar S, Periasamy A, Herman B, Juliano R 1991 Mechanism of cellular uptake of modified oligodeoxynucleotides containing methylphosphate linkages. Nucleic Acids Res 19:5543–5550

Sood S, Shaw RB, Spielvogel BF 1991 Boron-containing nucleic acids. Synthesis of oligonucleoside boranophosphates. J Am Chem Soc 112:9000–9001

Summerton J, Stein D, Huang SB, Matthews P, Weller D, Partridge M 1997 Morpholino and phosphorothioate antisense oligomers compared in cell-free and in-cell systems. Antisense Nucleic Acid Drug Dev 7:63–70

Van Aerschot A, Verheggen I, Hendrix C, Herdewijn P 1995 1,5-Anhydrohexitol nucleic acids, a new promising antisense construct. Angew Chem Int Ed Engl 34:1338–1340

Wagner RW, Matteucci MD, Lewis JG, Gutierrez AJ, Moulds C, Froehler BC 1993 Antisense gene inhibition by oligonucleotides containing C-5 propyne pyrimidines. Science 260:1510–1513

Wagner RW, Matteucci MD, Grant D, Huang T, Froehler BC 1996 Potent and selective inhibition of gene expression by an antisense heptanucleotide. Nat Biotechnol 14:840–844

Wyatt JR, Vickers TA, Roberson JL et al 1994 Combinatorially selected guanosine-quartet structure is a potent inhibitor of human immunodeficiency virus envelope-mediated cell fusion. Proc Natl Acad Sci USA 91:1356–1360

DISCUSSION

Gait: You gave a good overview of neutral backbone analogues, and I would first like to add to the historical overview by stating that this field of study originated in Birmingham in the late 1960s (Halford & Jones 1968). I was one of the graduate students involved there in the early 1970s. Secondly, I get the impression from your presentation that, in your opinion, we should ignore the neutral backbone analogues and concentrate on the sugar and base analogues.

Matteucci: No, my point was that most isosteres of phosphorus are difficult to make. Peptide nucleic acids (PNAs) or morpholinophosphorodiamidines are the exception, and therefore they could be more fruitful.

Gait: But is it possible to adjust all the parameters you need merely by using sugar and base analogues?

Matteucci: Yes, it may be possible. Many people are making chimeric molecules, i.e. mixing large regions of neutral molecules with a small amount of phosphorothioate in the middle which activates RNase H, and they are attempting to attach methylphosphonates, PNAs and 2'-methoxyethoxy groups, for example, to the neutral 'wings'. However, this involves a substantial amount of work, and we will probably end up using those molecules that work effectively and are the simplest to make.

Caruthers: But we shouldn't discount something just because it's difficult to make. If it works people will discover easier ways of making it.

Cohen: People are currently working with relatively simple phosphorothioates, but additional modifications increase the cost of the product and the time it takes to make it, so it is important to think about how much is actually gained relative to the additional costs.

Wickstrom: Examples do exist of effective drugs that are relatively insoluble and were originally difficult to develop. Steroids and prostaglandins, for example, are relatively insoluble, but they work at low doses and it's now possible to administer them pharmaceutically.

Iversen: I would like to comment on the morpholene compounds, which we have tested in a rat model (Desjardins & Iversen 1995). We found that they have favourable pharmacokinetics, and that they selectively suppress the expression of the gene encoding cytochrome P450 3AZ. This is intriguing because these compounds don't support RNase activation. The suspicion is that they don't enter cells *in vitro*, although this is probably not the case *in vivo*, at least in the liver.

Caruthers: Have you observed any immune activation when using these compounds?

Iversen: There wasn't any short-term activation of the immune system, although the permeation question is controversial and difficult to sort out.

Matteucci: How did you follow the pharmacokinetics of the morpholene derivatives?

Iversen: We put a fluorescein label on the 5′ end, annealed the derivatives to a complementary phosphodiester and then ran them through an Applied Biosystems (Foster City, CA) DNA sequencer.

Matteucci: You mentioned that the pharmacokinetics are favourable, but how favourable is this relative to phosphorothioates, for example?

Iversen: The most important factor is the volume of distribution. A typical value for general distribution is 1 l/kg, whereas these compounds have a distribution of 2.7–3 l/kg, indicating that they bind outside the bloodstream and that there is cellular internalization, or at least absorption to tissues outside the bloodstream. This volume distribution is similar to phosphorothioates but the dose required to suppress liver enzyme expression is much lower than for phosphorothioates. This may be due to the higher T_m.

Stein: I have three comments. Firstly, we have also had some success with phosphorothioates in a rat model of restenosis using the anti-c-*myc*-targeted oligonucleotide.

My second comment is that I have problems with discussing permeation in non-real time, i.e. in a situation where the oligonucleotide is given to an animal that is then sacrificed in order to obtain the tissue (compared to real time, where living tissue is being observed). There are major difficulties with extrapolating a real-time experiment to a non-real-time experiment.

Thirdly, in my opinion a universal oligonucleotide does not exist. There are too many differences among different cell types, different tissues and different species.

Extrapolations from one system to another are therefore not going to work. There are many modifications, which we will have to fine-tune to the system in which we are working.

Agrawal: Many antisense DNA analogues have been made, including phosphorothioates, methylphosphonates, and those with base and sugar modifications; however, not all of these analogues show biological activity in cell culture. What are the reasons for this? They have a high level of nuclease resistance and a high T_m, which are considered to be important factors responsible for the biological activity in the intact cell culture system. Phosphorothioates show sequence- and non-sequence-specific activity in cell culture, but only sequence-specific activity *in vivo*. What, therefore, are the differences between these two systems? To make phosphorothioates more effective as drugs, we need to mix and match these various analogues. However, there are still certain limitations in terms of their understanding of safety profiles and pharmacokinetics.

Caruthers: Have you observed changes in tissue distribution by attaching other functional moieties, in addition to sulfur, to non-RNase-active regions of the phosphorothioate?

Agrawal: The majority of the phosphorothioates have similar pharmacokinetic profiles, although significant differences have been observed in studies based on the use of ^{35}S-labelled oligonucleotides. We (Agrawal et al 1991) and others (Sands et al 1994, Iversen et al 1994) have observed that a major pathway of elimination is via urinary excretion, although studies based on the use of ^{14}C-labelled oligonucleotides have reported elimination via expired air (Cossum et al 1993), a mechanism which is not clearly established. Also, significant changes have been observed in the pharmacokinetics of phosphorothioates containing contiguous guanosine residues. These phosphorothioate oligonucleotides are known to form hyperstructures, and following intravenous administration, we have observed differences in plasma clearance, tissue disposition, elimination and *in vivo* stability.

Caruthers: Is it possible, therefore, that the sequence has a stronger influence on the distribution than the analogue backbone?

Agrawal: The distribution depends on a number of factors. For example, macrophages take up specific oligonucleotides and re-direct them to other compartments. Also, for phosphorothioate oligonucleotides attached to cholesterol, 60–80% of the injected dose is disposed to liver tissues within 30 min of intravenous administration. These oligonucleotides are trapped there, i.e. they are not re-distributed to other tissues. By placing nuclease-resistant analogues at the end of phosphorothioate oligonucleotides, we have been able to increase *in vivo* stability, both in plasma and tissues.

Monia: We have looked at PNAs in cell culture and found that they have lagged behind some of the other analogues with respect to activity in cells, primarily for two reasons: (1) they do not permeate the cell membrane readily, and so are not amenable to conventional transfection techniques; and (2) they do not support RNase H activity when duplexed with RNA, and therefore it is difficult to use them because novel (non-

RNase H) terminating mechanisms must be employed. Furthermore, it is difficult to make chimeric molecules with them that would permit RNase H activity. The problem with non-RNase H-terminating mechanisms is that it is difficult to find spots that will work through non-RNase H mechanisms of action, such as steric blocking, but when we have found such a site PNA oligomers do work relatively well in culture. They do not affect RNA levels but they are potent and specific inhibitors of protein translation, provided that they can be targeted to an optimal site of action, and decorating them with certain charged residues, such as lysine or aspartic acid, affects their ability to be transfected into cells efficiently. I'm not an expert on pharmacokinetics, but I know that they also have a different distribution to phosphorothioate oligonucleotides. Also, the PNAs that we have found to work well in cell culture are relatively short in length, i.e. 12–13 mer.

Caruthers: Were these PNAs taken up by cells using a liposome preparation?

Monia: No, either by electroporation, for example, or other methods that create holes in cell membranes. They don't work well with most of the cationic lipids that we've looked at.

Wickstrom: Mark Matteucci mentioned that phosphorothioate derivatives bind to proteins. Indeed, we will probably find that all DNA derivatives are capable of binding to proteins. However, we don't have the resources to characterize the K_d of all these DNA derivatives binding to all the proteins, and the best we may be able to do is to look out for toxic effects at high doses.

Gait: There have been many suggestions that all phosphorothioate oligonucleotides bind protein to a large extent, which is a problem. Is this merely a titration phenomenon, i.e. is there a cut-off for the number of phosphorothioates in an oligonucleotide below which suddenly there's no protein binding?

Stein: It is unlikely that there is a charged oligonucleotide that does not bind protein. It's probably a question of relative dissociation constants. It seems that the proteins to which oligonucleotides bind are predominantly heparin-binding proteins. I am not aware of any other classes of proteins that bind to oligonucleotides under physiological conditions. However, the average fall-off in the dissociation constant between an isosequential phosphorothioate and a phosphodiester is about one to three orders of magnitude. Therefore, if a phosphodiester binds with a micromolar dissociation constant then the dissociation constant of the corresponding phosphorothioate is in the nanomolar range. This is when one starts noticing biological effects because the K_d is approaching the K_d of the natural ligand. There is a certain length effect, although it is not quite linear: there seems to be a cut-off at about 15 phosphorothioate linkages, such that below 15 it falls off dramatically and at 22–24 linkages it plateaues. This behaviour is not just a function of phosphorothioates but also of the molecules that the phosphorothioates interact with. The reasons for the change in dissociation constants are not well understood at present.

Caruthers: Is it also the case that in your systems the dithioates are non-competitive?

Stein: No, although it depends on the protein. Fibroblast growth factor 2 and CD4 are competitive, for example, but it is not true in every case.

Agrawal: Binding of phosphorothioate oligonucleotides to serum and plasma protein is also dependent on the species of animal. It is difficult to say whether sulfur is responsible for protein binding because we have observed reduced protein binding in some phosphorothioate oligonucleotides that contain sulfur (e.g. phosphorothioate oligonucleotides or their 2'-O-methyl analogues). Binding to proteins also depends on the length of phosphorothioate oligonucleotide.

Gait: So the answer seems to be that the problems of protein binding and toxicity can be solved by titration of the number of phosphorothioate linkages.

Monia: Possibly, but a certain amount of protein binding will be desired pharmacokinetically.

Stein: Also, protein binding and internalization are intimately linked because a PNA that doesn't bind proteins well has a decreased rate of internalization. On the other hand, phosphorothioates bind proteins well and there are no internalization problems. Presumably there's a middle ground where it is possible to take the advantages of both.

Caruthers: Is there any evidence that phosphorothioates are not in this middle ground?

Stein: It depends both on the sequence (i.e. whether the molecule contains a number of guanine residues) and on the length of the oligonucleotide, because in terms of *in vivo* toxicity the longer the oligonucleotide and the more phosphorothioate residues there are, the more toxic the compound will be. However, if you decrease the number of phosphorothioate residues, then you may develop problems with internalization.

References

Agrawal S, Temsamani J, Tang J-Y 1991 Pharmacokinetics, biodistribution, and stability of oligodeoxynucleotide phosphorothioates in mice. Proc Natl Acad Sci USA 88:7595–7599

Cossum PA, Sasmor H, Dellinger D et al 1993 Disposition of the ^{14}C-labeled phosphorothioate oligonucleotides ISIS 2105 after intravenous administration to rats. J Pharmacol Exp Ther 267:1181–1190

Desjardins JP, Iversen PL 1995 Inhibition of the rat cytochrome P450 3AZ by an antisense phosphorothioate oligonucleotide *in vivo*. J Pharmacol Exp Ther 275:1608–1613

Halford MR, Jones AS 1968 Synthetic analogues of polynucleotides. Nature 217:638–640

Iversen PL, Mata J, Tracewell WG, Zon G 1994 Pharmacokinetics of an antisense phosphorothioate oligonucleotide against *rev* from human immunodeficiency virus type 1 in the adult male rat following single injections and continuous infusion. Antisense Res Dev 4:43–52

Sands H, Gorey-Feret LJ, Cocuzza AJ 1994 Biodistribution and metabolism of internally ^{3}H-labeled oligonucleotides. I. Comparison of a phosphodiester and a phosphorothioate. Mol Pharmacol 45:932–943

Phosphorothioate oligodeoxynucleotides: large-scale synthesis and analysis, impurity characterization, and the effects of phosphorus stereochemistry

Karen L. Fearon, Bernard L. Hirschbein, Choi-Ying Chiu, Maria R. Quijano and Gerald Zon

Chemical Process Research & Development, Lynx Therapeutics Inc., 3832 Bay Center Place, Hayward, CA 94545, USA

Abstract. Large-scale synthesis of phosphorothioate oligodeoxynucleotides on Tentagel™ using a 'batch mode' synthesizer and β-cyanoethyl phosphoramidite coupling followed by sulfurization with bis(O,O-diisopropoxy phosphinothioyl) disulfide (S-tetra™) provides stepwise yields of 98–99% and results in phosphorothioate oligodeoxynucleotides that are 93–97% pure, as determined by PAGE, after reverse-phase high performance liquid chromatography (RP-HPLC) and 'downstream' processing. The purity of phosphorothioate oligodeoxynucleotides synthesized on Tentagel™ is significantly higher than those synthesized on controlled pore glass. Electrospray ionization mass spectrometry of the $n-1$ impurity isolated by preparative PAGE was used to establish that the $n-1$ impurity is a heterogeneous mixture of all possible single-deletion sequences, relative to the parent phosphorothioate oligodeoxynucleotide, and results from minor, though repetitive, imperfections in the synthesis cycle. Acid-catalysed depurination was found to occur both during the synthesis and during the post-synthesis detritylation, following RP-HPLC. Studies of hybridization affinity and biological mechanism of action using independently synthesized $n-1$ phosphorothioate oligodeoxynucleotides relative to the 15 mer LR-3280 showed that, in this case, the majority of the $n-1$ sequences had more than a 10 °C decrease in melting temperature with sense RNA compared to the n-mer, and they did not cause detectable cleavage of RNA by RNase H in HL-60 human promyelocytic leukaemia cells. P stereoregular phosphorothioate oligodeoxynucleotides are not significantly more active than their stereorandom counterparts and thus their use in clinical studies seems unwarranted.

1997 Oligonucleotides as therapeutic agents. Wiley, Chichester (Ciba Foundation Symposium 209) p 19–37

The ability of oligonucleotides to hybridize selectively to a target RNA or DNA, and thereby control genetic expression, has generated a significant interest in their synthesis, purification and characterization (Uhlmann & Peyman 1990). First-generation 'antisense' agents, phosphorothioate oligodeoxynucleotides, in which a sulfur atom replaces one of the non-bridging oxygen atoms of the phosphodiester linkage, have now reached the clinic, thus intensifying the need for large quantities of highly pure, well-characterized phosphorothioate oligodeoxynucleotides. Phosphorothioate oligodeoxynucleotides are currently being evaluated in clinical trials for their ability to inhibit restenosis (Shi et al 1994), various cancers (Bishop et al 1996, Gewirtz et al 1996, BioWorld Today 1996) and viruses (Galbraith et al 1994). These therapeutic applications required process improvements in the chain-assembly chemistry, increases in the scale of the automated synthesis hardware, and optimization of large-scale purification and 'downstream' processing. Analytical techniques to address issues of product identification and purity analysis, identification of impurities, and reproducibility of synthesis and purification methods also needed to be developed (Kambhampati et al 1993).

Our approach to the large-scale synthesis of phosphorothioate oligodeoxynucleotides has been to optimize the automated chain-assembly chemistries, while simultaneously reducing the cost of the reagents, such that a highly homogeneous 5'-dimethoxytrityl (DMT)-protected product is synthesized, thus allowing the use of a single, high throughput, reverse-phase high performance liquid chromatography (RP-HPLC) method for purification (Zon 1990). We currently use a proprietary 'batch mode' vortexing synthesizer that operates at the 1–10 mmol scale, which is adequate for the production of all the 15 mer phosphorothioate oligodeoxynucleotide (LR-3280) necessary for our anti-restenosis clinical applications. The yields and purity of the phosphorothioate oligodeoxynucleotides improved with scale-up, and there is no apparent reason why, when necessary, this scale-up could not be further increased. β-Cyanoethyl phosphoramidite coupling used together with bis(O,O-diisopropoxy phosphinothioyl) disulfide (S-tetra™) for sulfurization provides stepwise yields in the 98–99% range, as determined by the dimethoxytrityl cation assay (Stec et al 1993). Currently, 2.5–3.5 equivalents of phosphoramidite monomer are used per cycle; however, with rigorous drying of the monomer solutions, 2.0 equivalents are expected to give adequate coupling. S-tetra™ is an extremely efficient sulfurizing agent, with an average stepwise sulfur incorporation efficiency of 99.8%. Additionally, S-tetra™ is straightforward to synthesize and purify, and is a small fraction of the cost of the Beaucage reagent, widely used on the research scale (Iyer et al 1990). Proprietary capping agents, which terminate unreacted 5'-hydroxyl groups, are used to improve the purity of the RP-HPLC-isolated product. Exposure to acid during the detritylation step is minimized by incorporating conductivity monitoring with feedback to determine when the detritylation is complete, thereby minimizing acid-catalysed depurination of the N[6]-benzoyl 2'-deoxyadenosines and improving product purity (Horn & Urdea 1988).

Controlled pore glass (CPG) is the most commonly used solid support for oligodeoxynucleotide synthesis; however, its expense and relatively low loading (25–40 μmol/g) make it less than optimal for scale-up. CPG also contains reactive sites that support chain elongation and lead to shorter, DMT-containing impurities not removable by RP-HPLC (McCollum & Andrus 1991, Temsamani et al 1995). The use of a high loaded poly(styrene)–poly(oxyethylene) support (Tentagel[TM]; 150–170 μmol/g) reduces the amount of solid support required to produce a given amount of phosphorothioate oligodeoxynucleotide by a factor of four to six, resulting in increased throughput and significantly reduced cost. Additional savings are also achieved because the high loading allows for a reduction in the amount of solvent used for washing, on a per gram basis of phosphorothioate oligodeoxy-nucleotide synthesized. Most importantly, the 'DMT-on' RP-HPLC-purified product obtained from synthesis on Tentagel[TM] is consistently higher in chain-length purity, as measured by PAGE, than that obtained from synthesis on CPG (Table 1), presumably due to a combination of the inertness of the support and the lack of steric hindrance in the swellable matrix leading to more complete reactions. Currently, phosphorothioate oligodeoxynucleotides synthesized on Tentagel[TM] are typically 93–97% pure by PAGE after RP-HPLC. The product is detritylated, diafiltered, precipitated from 1 M aqueous NaCl with ethanol, depyrogenated and sterile filtered before the optional lyophilization step. Recently 61 g of a 15 mer phosphorothioate (LR-3280, 5'-AACGTTGAGGGGCAT-3') was synthesized in an overall isolated yield of 52.5%. The final product is subjected to extensive analysis to determine its purity, identity, quality and potency. These analyses include PAGE, electrospray ionization mass spectrometry (ESI-MS), ^{31}P nuclear magnetic resonance (NMR) spectrometry, strong anion-exchange chromatography, ^{1}H NMR spectrometry, sequencing, endotoxin and bioburden determination, pH, melting temperature with sense RNA, per cent water content, and residual chloride determination. An example of the analytical results for a typical batch of LR-3280 is shown below.

(1) Purity determined by PAGE separation, visualization of the resultant bands with Stains-all[TM] and integration using laser-scanning densitometry

n-mer	96%
n–1	2.6%
n–2	0.8%
n–3	0.3%

(2) Mass spectrometry — found: 4881 Da; calculated: 4881.9 Da

(3) ^{31}P NMR — 99.8% phosphorothioate linkage; 0.2% phosphodiester linkage

(4) ^{1}H NMR — 0.09% residual organics

(5) Per cent water — 2.5%

(6) Endotoxin — <0.03 endotoxin units (EU)/ml (limit of detection)

(7) Bioburden — 0 colony-forming units (CFU)/ml

(8)   Per cent chloride   0.002%
(9)   pH   7.3

The characterization of the low levels of impurities remaining in the purified phosphorothioate oligodeoxynucleotide products is important from both a regulatory standpoint, as well as for efforts aimed at further refining the synthesis chemistry. The single nucleotide deletion (n–1) is generally the predominant impurity (1.5–3% on Tentagel™; 4–6% on CPG) in the phosphorothioate oligodeoxynucleotide product (n) as determined by PAGE, with progressively truncated oligomers being less abundant. ESI-MS of the n–1 impurity isolated by preparative PAGE was used to establish, for the first time, that the n–1 impurity is a heterogeneous mixture of all possible single-deletion sequences, relative to the parent phosphorothioate oligodeoxynucleotide (Fearon et al 1995). These deletion sequences result from incomplete detritylation, incomplete coupling, followed by incomplete capping or incomplete sulfurization. For example, the reconstructed mass spectrum of the n–1 impurity that was gel isolated from LR-3523, a 20 mer phosphorothioate with the sequence 5'-CCCTGCTCCCCCCTGGCTCC-3', contains the masses representing all three possible types of unique n–1 species (Fig. 1): LR-3532 missing one of the 13 2'-deoxycytidine 3'-thiophosphates (found 5901.2 Da, calculated 5901.7 Da); one of the four thymidine 3'-thiophosphates (found 5886.0 Da, calculated 5885.7 Da); or one of the three 2'-deoxyguanosine 3'-thiophosphates (found 5862.2 Da, calculated 5861.7 Da). The heterogeneity of the n–1 species and, by extension, other shorter impurities leads to the conclusion that each possible 'short-mer' sequence is present in only a low abundance. Knowing the identity of these impurities allows for the systematic study of new or improved synthesis and purification procedures and also fulfils the requirements set forth by the Food and

TABLE 1   Comparison of the purity[a] of phosphorothioate oligodeoxynucleotides[b] synthesized on Tentagel™ and controlled pore glass

	Tentagel^TM				*Controlled pore glass*			
Sequence	n	n−1	n−2	n−3	n	n−1	n−2	n−3
5'-CGCTGAAGGGCTTCTTCCTTATTGAT-3'	94	2.8	1.2	0.2	90	4.4	2.5	1.5
5'-TATGCTGTGCCGGGGTCTTCGGGC-3'	94	2.5	1.0	0.4	87	5.9	2.6	1.1
5'-CCCTGCTCCCCCCTGGCTCC-3'	95	2.0	0.6	0.4	87	5.0	1.6	1.3
5'-CGCTGAAGGGCTTCTGCGTCTCCATG-3'	94	1.4	2.2	1.0	86	5.0	3.6	1.3

[a]Purity is determined by sequential PAGE separation, visualization of the resultant bands with Stains-all™ and integration using laser-scanning densitometry.
[b]Phosphorothioate oligodeoxynucleotides were purified by reverse-phase high performance liquid chromatography, detritylated, extracted and precipitated three times from 1 M aqueous NaCl with ethanol.

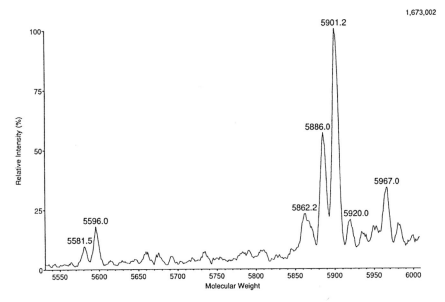

FIG. 1. Transformed mass spectrum of the PAGE-purified n–1 impurity from LR-3523, which also contains about 15% of the n–2 impurity. The three possible n–1 impurities are identified as LR-3523 missing a 2′-dC 3′-thiophosphate (found 5901.2 Da, calculated 5901.7 Da); a T 3′-thiophosphate (found 5886.0 Da, calculated 5885.7 Da); or a 2′-dG 3′-thiophosphate (found 5862.2 Da, calculated 5861.7 Da). Also present are two types of n–2 impurities: LR-3523 missing two 2′-dC 3′- thiophosphates (found 5596.0 Da, calculated 5596.5 Da); or one 2′-dC 3′-thiophosphate and one T 3′-thiophosphate (found 5581.5 Da, calculated 5581.5 Da). The peak at 5967.0 Da is an adduct of unknown identity. Figure 1 is reproduced from Fearon et al (1995) with the permission of Oxford University Press.

Drug Administration for complete characterization of related structural impurities for all drugs.

Acid-catalysed depurination of N^6-benzoyl 2′-deoxyadenosines during synthesis followed by ammoniolytic cleavage is the only side reaction that creates a relatively significant amount of a particular short-mer impurity (Horn & Urdea 1988). In the case of LR-3280, the deoxyadenosine moiety in the second position from the 3′ end necessarily has the most exposure to acid, relative to all other deoxyadenosine moieties in this sequence, because of the 3′ to 5′ direction of synthesis. Depurination at this site, followed by cleavage, leaves a DMT-containing impurity (Fig. 2) that co-elutes with the DMT-protected product during RP-HPLC. This impurity is seen in the mass spectrum of the LR-3280 product (Fig. 3) and the molecular weight (found 4444.0 Da) indicates that either water (calculated 4443.6 Da) or ammonia (calculated 4442.6 Da) added to the α,β-unsaturated imine formed upon loss of the adenine base. This impurity, which was independently synthesized using a t-butyldimethylsilyl-protected apurinic nucleoside monomer (Groebke & Leumann 1990), co-migrates

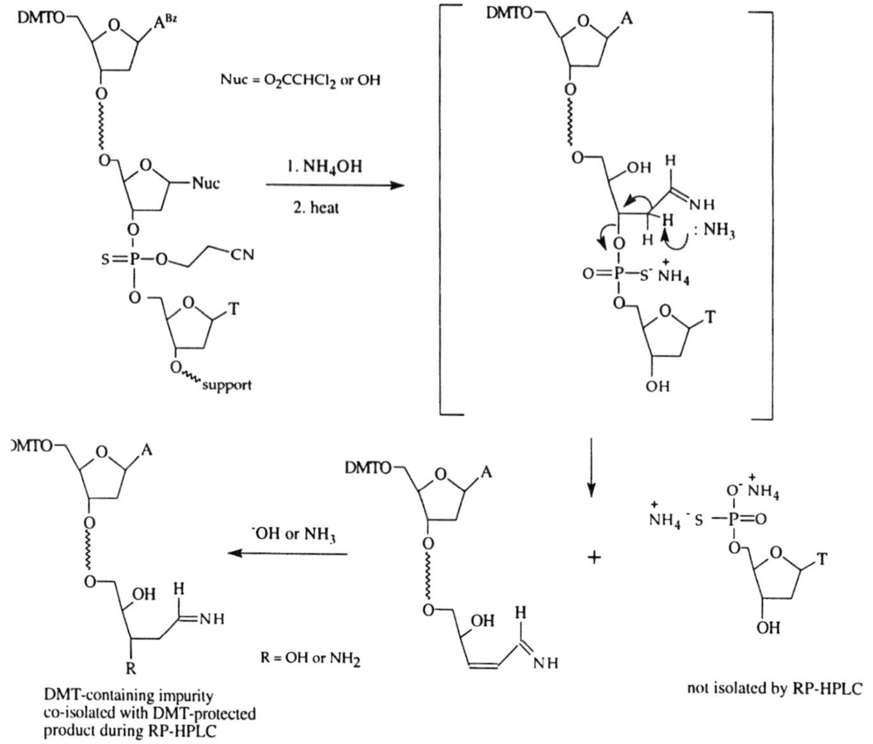

FIG. 2. Example of 5'-dimethoxytrityl (DMT)-containing impurity formation by a depurination/cleavage mechanism. RP-HPLC, reverse-phase high performance liquid chromatography. Nuc, O_2CCHCl_2 or OH.

with the n–1 impurities in LR-3280 under denaturing PAGE conditions. Depurination can be kept to a minimum by reducing the exposure of the oligonucleotides to acid by using conductivity monitoring with feedback and/or by changing the N^6 protection on deoxyadenosines from benzoyl to dialkylformamidine (Froehler & Matteucci 1983).

The finding that depurination also occurred during the post-synthesis detritylation following RP-HPLC was more surprising. These full-length apurinic impurities are not cleaved during subsequent downstream processing in which there is no treatment with base. In the case of LR-3280 (Fig. 3), the major apurinic impurity corresponds to the LR-3280 sequence with one A base replaced with a hydroxyl group, i.e. an A-minus apurinic site (found 4765.5 Da, calculated 4764.8 Da), while the relatively less abundant impurity is the analogous G-minus apurinic site (found 4749.5 Da, calculated 4748.8 Da). Furthermore, PAGE of LR-3280 and several different phosphorothioate oligodeoxynucleotide products showed that, under denaturing conditions, these impurities can migrate either as an n-like or n–1-like

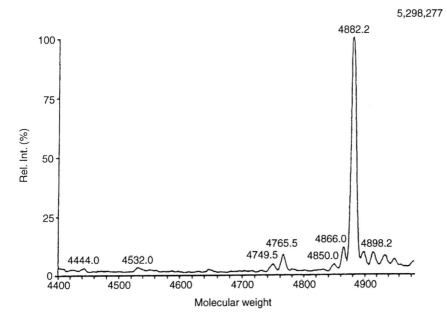

FIG. 3.   Transformed mass spectrum of the LR-3280 product (found 4882.2 Da, calculated 4881.9 Da). The single and double phosphodiester defect impurities of LR-3280 are observed at 4866.0 Da (calculated 4865.8 Da) and 4850.0 Da (calculated 4849.8 Da), respectively. The full-length A-minus (found 4765.5 Da, calculated 4764.8 Da) and G-minus (found 4749.5 Da, calculated 4748.8 Da) apurinic impurities are also identified.

species. ESI-MS and PAGE were important analytical tools for monitoring improved detritylation protocols, which eventually eliminated the formation of these full-length apurinic impurities.

The aforementioned characterization of the $n-1$ impurity allowed our use of independently synthesized $n-1$ phosphorothioate oligodeoxynucleotides as readily available and reasonable models of the authentic low-level $n-1$ impurities for studies of physical properties and biological mechanism of action necessary in preclinical investigations of phosphorothioate oligodeoxynucleotides. The melting temperatures of each of the synthetic $n-1$ sequences (LR-4276 to LR-4285) relative to LR3280 (n) with sense RNA were determined. This type of hybridization involves binding each $n-1$ phosphorothioate oligodeoxynucleotide to its target through 'bulged RNA'. The melting temperatures of the duplexes with single-base bulges are 1.2 °C to 17.9 °C lower (ΔT_m) than the melting temperature of the fully complementary duplex ($T_m = 50.9$ °C, Table 2). In this case, all base-bulged duplexes with the bulge more than three bases from the end of the duplex have a ΔT_m greater than 10 °C. The hybridization affinity and ability of these synthetic '$n-1$' phosphorothioate oligodeoxynucleotides to activate an RNase H cleavage mechanism in cell culture

Base Pairs

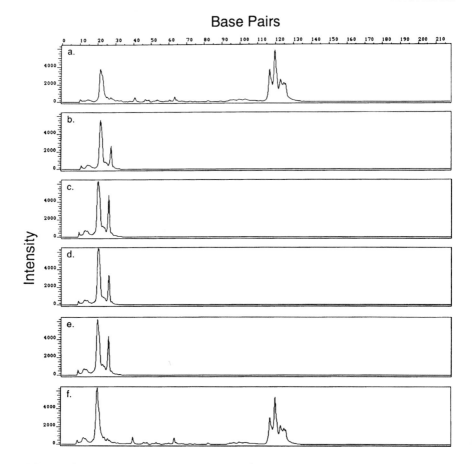

FIG. 4.   Electrophoretogram from PAGE of the reverse ligation-mediated PCR products from treatment of HL-60 human promyelocytic leukaemia cells for 24 h with 5 μM of: (a) LR-3280 (positive control); (b) LR-3001 (negative control); (c) LR-4278; (d) LR-4282; (e) LR-4283; and (f) LR-4284. Only LR-3280 and LR-4284 show PCR products due to cleavage of the *myc* message under the LR-3280 binding site: 113–127 bp. Intensity in arbitrary units.

was tested relative to the parent, LR-3280, using a reverse ligation-mediated (RL)-PCR assay (Giles et al 1995). HL-60 human promyelocytic leukaemia cells in log-phase growth were incubated at 37 °C with 5 μM phosphorothioate oligodeoxynucleotide for 24 h, followed by isolation of the RNA and RL-PCR as previously reported (Giles et al 1995). If the bulge was more than three bases from the end of the duplex, the *myc* RNA was not cleaved; however, those duplexes with the bulge near the end did cause cleavage of the *myc* message under the LR-3280 binding site (Fig. 4). It is not known whether the enzyme was unable to bind to the bulged duplex or cut the bulged RNA in the bound duplex, or if the $n{-}1$ phosphorothioate

TABLE 2 **Relative melting temperatures for LR-3280 and related $n-1$ sequences with sense RNA**

Name	Sequence	$(Tm)^a$ or ΔTm^b $(°C)$
LR-3280	Antisense phosphorothioate oligodeoxynucleotide 5'-AACGTTGAGGGGCAT-3' 3'-UUGCAACUCCCCGUA-5' Sense phosphodiester RNA	$(50.9)^c$
Bulged duplexes LR-4276	5'---ACGTTGAGGGGCAT-3' U- U	3.0
LR-4277	5'-AA--GTTGAGGGGCAT-3' -U C G	2.5
LR-4278	5'-AAC--TTGAGGGGCAT-3' -G A- C	17.9
LR-4279	5'-AACG--TGAGGGCAT-3' -C A- A	11.1
LR-4280	5'-AACGTT--AGGGGCAT-3' -A U- C	17.9
LR-4281	5'-AACGTTG--GGGGCAT-3' -C C- U	13.8
LR-4282	5'-AACGTTGA--GGGCAT-3' -U C- C	15.2
LR-4283	5'-AACGTTGAGGGG--AT-3' -C U- G	12.0
LR-4284	5'-AACGTTGAGGGGC--T-3' -G A U	1.2
LR-4285	5'-AACGTTGAGGGGCA -3' -U A	2.6

[a] T_m experiments were performed in PIPES buffer with 0.1 M NaCl and 3.5×10^{-6} M strand concentration.

[b] ΔT_m is defined as the T_m of the complexes with antisense strands having deletions subtracted from the T_m of the corresponding unmodified complex.

[c] The T_m value of the unmodified duplex is in parentheses.

oligodeoxynucleotides with base deletions more than three bases from the ends did not hybridize to the RNA in cell culture due to their lower hybridization affinities. Regardless, the low concentration of the individual $n-1$ impurities in the LR-3280 product and the lack of activity for most of them indicate that the n-mer itself is responsible for the observed efficacy in the porcine model (Shi et al 1994). Perhaps more importantly, it is unlikely that any of the individual $n-1$ sequences have biological activity against any target due to their extremely low concentration.

The phosphorothioate oligodeoxynucleotides synthesized using phosphoramidite chemistry followed by sulfurization are a mixture of P-chiral diastereomers wherein the number of diastereomers for a sequence with n linkages is equal to 2^n. There is widespread interest in stereopure phosphorothioate oligodeoxynucleotides, and with the recent development of the oxathiaphospholane chemistry, uniformly stereocontrolled phosphorothioate oligodeoxynucleotides are now attainable (Stec et al 1995). We synthesized the all-R_p and all-S_p versions of LR-3280, as well as other sequences, in order to determine their relative hybridization efficiencies with RNA and biological effects $in\ vitro$. The relative order of melting temperatures for the LR-3280-related compounds with sense RNA follows the trend previously reported (Koziolkiewicz et al 1995, Tang et al 1995) and also found for other sequences in our laboratory, and is all-R_p ($T_m = 55.2\,°C$) > stereorandom ($T_m = 50.7\,°C$) > all-S_p ($T_m = 48.3\,°C$). The activity of these compounds in $in\ vitro$ studies, including the inhibition of proliferation of low passage, serum-stimulated vascular smooth muscle cells (Shi et al 1994) by the LR-3280 analogues, shows the same trend: all R_p > stereorandom > all-S_p (Fig. 5). The greater efficacy of the R_p analogues, relative to their S_p counterparts, is most likely a result of their increased hybridization efficiency and their superior ability to activate RNase H, rather than resistance to degradation by serum nucleases, which are R_p diastereoselective (Koziolkiewicz et al 1995). Although the all-R_p phosphorothioate oligodeoxynucleotides show consistent, reproducible biological effects that are greater than the all-S_p phosphorothioate oligo-deoxynucleotides, they are not significantly more active than the much less expensive stereorandom phosphorothioate oligodeoxynucleotides, and thus their use in clinical settings seems unwarranted.

Phosphorothioate oligodeoxynucleotides have led the way toward proving that 'antisense' therapeutics can be new, useful and cost-effective treatments for at least some clinical indications. However, phosphorothioate oligodeoxynucleotides sometimes have non-specific effects presumably due to adventitious protein binding. Second-generation analogues, such as the N3′→P5′ phosphoramidate oligonucleotides, have improved properties for antisense and antigene therapies, as well as research and diagnostic applications. Uniformly modified N3′→P5′ phosphoramidate oligonucleotides, wherein each 3′ oxygen is replaced by a 3′ amine, have a high affinity for complementary single-stranded RNA and single- and double-stranded DNA, do not bind non-specifically to proteins, and are resistant to enzyme degradation (Chen et al 1995, Gryaznov et al 1995). Moreover, a 15 mer N3′→P5′ phosphoramidate oligonucleotide sequence showed efficacy $in\ vivo$ against chronic

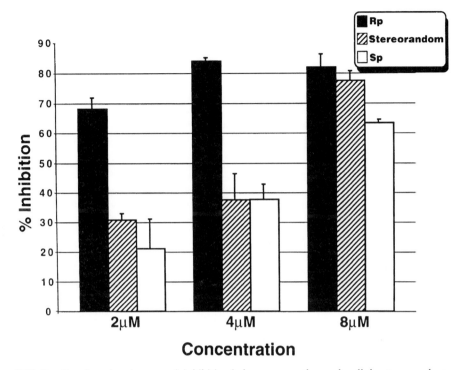

FIG. 5.   Bar chart showing growth inhibition in human smooth muscle cells by stereorandom, R_p and S_p LR-3280 at concentrations of 2, 4, and 8 μM.

myelogenous leukaemia at one-tenth the comparably effective dose of the isosequential phosphorothioate oligodeoxynucleotide (Skorski et al 1997). In addition, another 15 mer N3′→P5′ phosphoramidate oligonucleotide sequence showed no toxicity in preliminary *in vivo* studies in mice at doses up to 150 mg/kg administered intravenously six times over two weeks, whereas kidney and liver toxicity was seen for the isosequential phosphorothioate oligodeoxynucleotide at the same dose (C. Gamba-Vitelo & G. Zon, unpublished results 1995). We are aggressively pursuing the process development, manufacture and evaluation of the biological effects of these promising second-generation analogues (Gryaznov et al 1996, McCurdy et al 1996), as well as continuing our work with phosphorothioate oligodeoxynucleotides.

Acknowledgements

We thank Colleen Hansen for the analysis of the clinical lot of LR-3280, John Stults for performing the ESI-MS on the phosphorothioate oligodeoxynucleotide products and impurities, Lynda Ratmeyer for melting temperature studies, Wojciech Stec for the oxathiaphospholane monomers and Yi Shi for the *in vitro* studies of the stereopure LR-3280 compounds in vascular smooth muscle cells. We also thank Pamela Dal Bozzo and Frederic

Olivieri for help with the figure preparations. The stereopure phosphorothioate oligodeoxynucleotide project was funded by SBIR grant no. 2R44 AI34663-02.

References

BioWorld Today 1996 Isis, Ciba finalize collaboration, take first compound into clinic. 7:1–2

Bishop MR, Iversen PL, Bayever E et al 1996 Phase I trial of an antisense oligonucleotide OL(1)p53 in hematologic malignancies. J Clin Oncol 14:1320–1326

Chen J-K, Schultz RG, Lloyd DH, Gryaznov SM 1995 Synthesis of oligodeoxyribonucleotide N3'→P5' phosphoramidates. Nucleic Acids Res 23:2661–2668

Fearon KL, Stults JT, Bergot BJ, Christenesen LM, Raible AM 1995 Investigation of the 'n–1' impurity in phosphorothioate oligodeoxynucleotides synthesized by the solid-phase β-cyanoethyl phosphoramidite method using stepwise sulfurization. Nucleic Acids Res 23:2754–2761

Froehler BC, Matteucci MD 1983 Dialkylformamidines: depurination resistant N^6-protecting group for deoxyadenosine. Nucleic Acids Res 11:8031–8036

Galbraith WM, Hobson WC, Giclas RC, Schechter PJ, Agrawal S 1994 Complement activation and hemodynamic changes following intravenous administration of phosphorothioate oligonucleotides in the monkey. Antisense Res Dev 4:201–206

Gewirtz AW, Luger S, Sokol D et al 1996 Treating human myelogenous leukemia with c-*myb* antisense oligodeoxynucleotides: two years clinical experience. J Invest Med 44:279A

Giles RV, Spiller DG, Tidd DM 1995 Detection of ribonuclease H-generated mRNA fragments in human leukemia cells following reversible membrane permeabilization in the presence of antisense oligodeoxynucleotides. Antisense Res Dev 5:23–31

Groebke K, Leumann C 1990 A method for preparing oligodeoxynucleotides containing an apurinic site. Helv Chim Acta 73:608–617

Gryaznov SM, Lloyd DH, Chen J-K et al 1995 Oligonucleotide N3'→P5' phosphoramidates. Proc Natl Acad Sci USA 92:5798–5802

Gryaznov S, Skorski T, Cucco C et al 1996 Oligonucleotide N3'→P5' phosphoramidates as antisense agents. Nucleic Acids Res 24:1508–1514

Horn T, Urdea MS 1988 Solid supported hydrolysis of apurinic sites in synthetic oligonucleotides for rapid and efficient purification on reverse-phase cartridges. Nucleic Acids Res 16:11559–11571

Iyer RP, Phillips LR, Egan W, Regan JB, Beaucage SL 1990 The automated synthesis of sulfur-containing oligodeoxyribonucleotides using 3-*H*-1,2-benzodithiol-3-one 1,1-dioxide as a sulfur-transfer reagent. J Org Chem 55:4693–4699

Kambhampati RVB, Chiu Y-Y, Chen CW, Blumenstein JJ 1993 Regulatory concerns for the chemistry, manufacturing, and controls of oligonucleotide therapeutics for use in clinical studies. Antisense Res Dev 3:405–410

Koziolkiewicz M, Krakowiak A, Kwinkowski M, Boczkowska M, Stec WJ 1995 Stereodifferentiation — the effect of P chirality of oligo(nucleoside phosphorothioates) on the activity of bacterial RNase H. Nucleic Acids Res 23:5000–5005

McCollum C, Andrus A 1991 An optimized polystyrene support for rapid, efficient oligonucleotide synthesis. Tetrahedron Lett 32:4069–4072

McCurdy SN, Nelson JS, Hirschbein BL, Fearon KL 1997 An improved method for the synthesis of N3'→P5' phosphoramidate oligonucleotides. Tetrahedron Lett 38:207–210

Ratajczak MZ, Kant JA, Luger SM et al 1992 *In vivo* treatment of human leukemia in a scid mouse model with c-*myb* antisense oligodeoxynucleotides. Proc Natl Acad Sci USA 89:11823–11827

Shi Y, Fard A, Galeo A et al 1994 Transcatheter delivery of c-*myc* antisense oligomers reduced neointimal formation in a porcine model of coronary artery balloon injury. Circulation 90:944–951

Skorski T, Perrotti D, Nieborowska-Skorska M, Gryaznov S, Calabretta B 1997 Antileukemic effect of c-*myc* N3'→P5' phosphoramidate antisense oligonucleotide *in vivo*. Proc Natl Acad Sci USA 94:3966–3971

Stec WJ, Uznanski B, Wilk A, Hirschbein BL, Fearon KL, Bergot BJ 1993 *Bis*(O,O–diisopropoxy phosphinothioyl) disulfide: a highly efficient sulfurizing reagent for cost-effective synthesis of oligo(nucleoside phosphorothioate)s. Tetrahedron Lett 34:5317–5320

Stec WJ, Grajkowski A, Kobylanska A et al 1995 Diastereomers of nucleoside 3'-O-(2-thio-1,3,2-oxathia(selena)phospholanes): building blocks for stereocontrolled synthesis of oligo(nucleoside phosphorothioate)s. J Am Chem Soc 117:12019–12029

Tang J, Roskey A, Li Y, Agrawal S 1995 Enzymatic synthesis of stereoregular (all R_p) oligonucleotide phosphorothioate and its properties. Nucleosides Nucleotides 14:985–990

Temsamani J, Kubert M, Agrawal S 1995 Sequence identity of the *n*–1 product of a synthetic oligonucleotide. Nucleic Acids Res 23:1841–1844

Uhlmann E, Peyman A 1990 Antisense oligonucleotides: a new therapeutic principle. Chem Rev 90:543–584

Zon G 1990 Purification of synthetic oligodeoxyribonucleotides. In: Hancock WS (ed) High performance liquid chromatography in biotechnology. Wiley, New York, p 301–397

DISCUSSION

Lebleu: Fibroblast growth factor (FGF) 2 is an essential growth factor, and since it has been suggested that phosphorothioate oligonucleotides are inhibitors of FGF2 action by interacting with its receptor, I was wondering whether you tested the R_p and S_p stereoisomers of this phosphorothioate oligonucleotide in this capacity?

Fearon: No, we have not.

Stein: We have just published a paper on this (Benimetskaya et al 1995). We didn't observe any differences in affinity of either the all-R_p or all-S_p isosequential forms of the phosphorothioates for FGF2 or for a number of other proteins that we tested.

I have another question relating to the potential problems of some of the G_4-containing oligonucleotides. What did you do to rule out the possibility that higher-order structures were present in your oligonucleotides?

Fearon: Higher-order structures are present in these oligonucleotides. We see both tetraplex and monomer by size-exclusion chromatography. This has been a thorn in our side, but we can show that an antisense mechanism is operating, although it may not be the only mechanism working. The cell culture reverse-ligation (RL)-PCR data show that the RNA is cleaved specifically, and in the *in vivo* pig studies there was no statistically significant effect in the G_4 control at the concentrations studied.

Stein: What was your G_4 control?

Fearon: It was a seven base mismatch that still contained G_4 in the same region as LR-3280 and it had the same base composition.

Stein: At the risk of being called a sceptic, I have to say that Maltese et al (1995) have shown that there are contextual differences in the antisense efficacy of a G_4-containing oligonucleotide. This 'G$_4$-ness' is strange and complicated because the rates at which these tetraplexes form (and the formation of higher-order structures is not reserved only for intramolecular tetrads) depend on many things, including the sequence, the position of the G$_4$, the ionic strength and even how long it is kept in the freezer.

Fearon: I agree, but we do know that this particular control has similar properties to the LR-3280 oligonucleotide. Moreover, when we dissolve the lyophilized LR-3280 product we reproducibly observe a 60:40 ratio of monomer:tetraplex, over a wide range of concentrations (about 10–10 000 μM) and for all batches of lyophilized LR-3280, regardless of how long it has been stored at either 5 °C or 25 °C. We feel that since we're able to show a biological effect *in vivo* of the compound against restenosis without toxicity then, regardless of exactly how it's working, we should develop it.

Caruthers: Did you look at the effect of single base pair mismatches of the G$_4$ region?

Fearon: Yes. If there's a mismatch in the G$_4$ region there's no activity.

Cohen: Several years ago we published a paper in which we showed that an oligomer with a G$_4$ sequence had a biological effect (Yaswen et al 1993). We made point mutations in this sequence and showed that the effect was due to the presence of this G$_4$ sequence, and that it was not an antisense effect. This effect was extremely specific, i.e. it was cell type specific and specific to phosphorothioates; therefore, our conclusion was that there is probably a receptor to which the G$_4$ phosphorothioate binds to give a more rapid biological effect than that which is observed as a result of antisense effects.

Wickstrom: It is interesting that the stereochemistry doesn't have as much of an impact as one might have thought, and it is encouraging that the tetraplex falls apart at 37 °C. We haven't done single nucleotide changes in the G$_4$ region, but if we take out the CT that lies between the G$_4$ and the CG motif we don't observe a reduction in Myc antigen.

Akhtar: We have an oligonucleotide that inhibits the tyrosine kinase activity of a receptor, and we found that the GGGAGG region within this oligonucleotide is responsible because when we substitute a single G with a T we lose the activity. The activity is also dependent on the chemistry because a phosphorothioate is active but a phosphodiester or a 2′-O-methyl-modified sequence is not.

Iversen: In the R$_p$ versus S$_p$ smooth muscle assay 2 μM of R$_p$ didn't seem to result in a robust difference, but there was a specific response at 4 μM and 8 μM. Is it possible that the R$_p$ inhibits RNase H more strongly than S$_p$?

Fearon: The differences between the activity of the R$_p$ and S$_p$ LR-3280 sequences in the smooth muscle cell assay were actually the most pronounced at 2 μM, with the R$_p$ LR-3280 analogue being the most active. We have not studied the inhibition of RNase H by the stereopure phosphorothioates; however, it has been reported (Koziolkiewicz et al 1995) that R$_p$ phosphorothioates are more susceptible to RNase H-dependent degradation compared to their stereorandom or S$_p$ counterparts and this may explain the higher activity of the R$_p$ analogues.

Ohtsuka: Is it possible that the stereopure oligonucleotides are more advantageous because they can be used at lower concentrations?

Fearon: In my opinion, they don't show enough improved activity to justify the significant increases in cost. But that doesn't mean we're not going to continue to study them. We are planning to do *in vivo* efficacy studies and compare their toxic effects with stereorandom oligonucleotides. However, in light of Cy Stein's observations (Benimetskaya et al 1995) that their protein-binding affinities are not dramatically different, and if their toxicity is caused mostly by adventitious protein binding, then we are unlikely to find differences in toxicity.

Eckstein: When looking at S_p and R_p phosphorothioate diastereomers, one should not forget the position of the phosphorothioate and its configuration, because this has a profound effect on the structure of the DNA, in terms of how easily it goes from the B to the Z conformation. This indicates that a structural modulation is occurring, which we really don't yet understand.

Has the Food and Drug Administration (FDA) approved the use of oligonucleotide stereoisomeric mixtures for clinical trials?

Fearon: Yes, they seem willing to accept stereorandom oligonucleotides. Isis has nearly obtained approval for using a particular stereorandom phosphorothioate.

Eckstein: You showed that the full-length *n*-mer is biologically active by RL-PCR and by showing that the RNA was cleaved at the expected positions. Did you show that the *n*–1 form does not have biological activity, thus ruling out the possibility that the *n*–1 form has RNase H-independent biological activity?

Fearon: If the *n*–1 sequence contains an intact G_4 region then it inhibits the growth of HL-60 human promyelocytic leukaemia cells, suggesting that it also has RNase H-independent biological activity.

Ohtsuka: I was also interested in your *n*–1 impurity because it gives a bulge in the heteroduplex. This may result in it binding more strongly to proteins, and thus increased toxicity.

Fearon: But the amount of *n*–1 impurity present in the product is small, therefore the likelihood of it causing a problem is low, even if we use a high concentration of oligonucleotide. If there is observable toxicity, it's probably the *n*-mer that's causing it.

Agrawal: The *n*–1 product is not a homogeneous product: we showed that the nucleotide may be missing from any part of the sequence (Temsamani et al 1995). It is possible that the *n*–1 product is generated by inefficient coupling and capping, but it is difficult to rule out other factors. In terms of whether this *n*–1 product or the *n*-mer is responsible for toxicity, the *n*–1 product only represents 2–10% of the total product, so in my opinion it should not add significantly to the toxicity of the parent *n*-mer. *n*–1 product is present in all batches of oligonucleotides used in each toxicity study that precedes each clinical study.

Caruthers: Is the presence of *n*–1 product, therefore, an important issue?

Agrawal: No. In the past the FDA has accepted up to 87% *n*-mer product, with the remainder composed of *n*–1, *n*–2 and *n*–3 species. Their focus has been on improving the quality. Presently, all the batches in use contain up to 90–95% of *n*-mer parent product.

I would also like to state that the differences between S_p and R_p also depend on experimental design. For example, we have shown that the R_p isomer has a higher T_m and is a better substrate for RNase H compared to the S_p isomer, but the R_p isomer is more nuclease sensitive than the S_p isomer (Tang et al 1995). Therefore, the

R_p isomer may have an increased ability to form a G_4 structure compared to the S_p isomer. This needs to be checked.

Fearon: We looked at that using size-exclusion chromatography and, at least in the case of LR-3280, we didn't see large differences between the amounts of quadruplex in the R_p versus S_p; it was approximately 60 : 40 monomer : tetramer in each case.

Stein: We also did not observe significant differences between the R_p and S_p in terms of the formation of higher-order structures.

Matteucci: I suspect that roughly half of the people in this room believe that this is working by antisense mechanism and half believe it is working via some sort of protein interaction. Your RL-PCR analysis suggests that there is an antisense effect in these HL-60 cells; however, it is possible that artefacts are generated during the RL-PCR assay. Our experience is that it is unusual to observe a cleaved product and to get an antisense result without any means of artificial delivery in cell culture.

Fearon: The cells are thoroughly washed to get rid of the oligonucleotide before the RNA is isolated.

Matteucci: But it could still be inside the endosomes.

Fearon: Yes, and that's an essential problem with that assay. Also, even if we do prove that an RNase H mechanism is operating, it doesn't mean that it's the only mechanism. Nevertheless, negative control experiments were conducted using detergent lysis of a 1 : 1 mixture of c-*myc*-expressing HL-60 cells, which had not been incubated with LR-3280, and a non-c-*myc*-expressing cell line, which had been incubated with LR-3280.

Wickstrom: It's inevitable that in cell culture assays if you observe sequence-specific c-*myc* mRNA truncation and you calculate that the K_d of the phosphorothioate binding to FGF allows an aptameric mechanism, then both of those mechanisms will occur in animals. There will also be other mechanisms operating. Nearly every drug that has been given to animals has been found to have multiple mechanisms of action. The FDA doesn't demand that we demonstrate that each drug has only a single mechanism of action, only that it is not too toxic. We originally did mismatch controls on the human c-*myc* codon 1–5 sequence in Karin Mölling's laboratory at the Max Planck Institut für Molekulare Genetik in Berlin (Wickstrom et al 1988). MC29 cells are quail cells transformed with an avian v-Myc oncogene, which has four mismatches in the codon 1–5 region equivalent to our target in human c-*myc*. These avian v-*myc*-transformed cells were not inhibited by the human anti-c-Myc DNA in the proliferation assay, implying that the human c-*myc* codon 1–5 antisense DNA had no activity against avian v-*myc*. Therefore, that is a good control in that the DNA drug of interest was tested in a different system where c-*myc* oncogene has four mismatches.

Matteucci: Did you try the converse, i.e. did you take the avian oligonucleotide sequence and target the avian c-*myc*?

Wickstrom: No, we didn't try that. We did those experiments about 12 years ago, but it's only recently that we've become interested in looking at them again. It would be a good control experiment to do. In my lab we have concentrated on sequences that don't have a G_4 sequence and are away from the first five codons, and we have found a mouse equivalent (MYC5) of the 5'-dTTTCATTGTTTTCCA-3' sequence that Gryaznov (1996) picked out near the C-terminus of c-*myc*. This sequence is more effective than our MYC6 (LR-3280) sequence and it doesn't have the G_4 sequence. This approach is easier than re-testing the MYC6 codon 1–5 sequence, although our studies with MYC6, which are now in Phase II, should be pursued because there are already abundant data on it.

Fearon: The practical view is that if it works well and there is no toxicity, and especially if no alternative therapy is available, then we should pursue it as a therapeutic agent, in parallel with continued mechanistic studies.

Hélène: We have also been concerned about the problem of artefacts using the RL-PCR technique. We did a series of experiments in which we added labelled RNA to the cell culture at the beginning of the purification procedure. We found that the labelled RNA was cleaved at the expected position, which means that one has to be careful when using the RL-PCR technique to determine whether the RNA is cleaved within cells. We have improved the purification procedure so at least we now know that under those conditions we are not cleaving an exogenous RNA in the cell culture system.

Iversen: To add to that, we've looked at the same assay for *p53* using an early time point at 2 h as a control. When we analysed RL-PCR samples using confocal microscopy we saw a diffuse pattern of the oligonucleotide in the nucleus. If the oligonucleotides were trapped inside endosomes as a result of receptor-mediated endocytosis and liberated from the endosomes so that they were available for cutting, one would expect to observe the cleaved product at 2 h as well as at 4 h. However, we only observed the cleaved product at 4 h. These results were dose dependent and sequence specific.

Krieg: I'm not sure that your controls exclude Claude Hélène's interpretation because with increasing time points there will be more oligonucleotide inside the cell, which explains why cleaved product might be seen only after 2 h. Similarly, in the dose experiments the higher the level of oligonucleotide inside the cell the more cleavage will be observed, and the cleavage should be sequence specific, as was observed.

Eckstein: I would like to address another point. What salt did you use in your oligonucleotide solution and have you observed different biological activities with other salts?

Fearon: We have always used sodium salt in our oligonucleotide solution, and we add a small amount of base to raise the pH to just over 7. We haven't tried using any other salts, although if we leave the oligonucleotides in triethylammonium acetate after reverse-phase high performance liquid chromatography we do observe some cellular toxicity.

Agrawal: We've tried using ammonium salts, but we found that this gives us problems with solubility. It is also possible to use potassium salts — they do not give rise to any toxic effects — but the oligonucleotides are less soluble as ammonium or potassium salts than as sodium salts.

Monia: I have a comment about the target sequence. c-*myc* is known to be regulated differentially as a function of proliferation state, and it also generates short RNA products when expression is reduced due to a termination site at the first exon following transcription. It is important to keep in mind, therefore, that when you're working with such regulated transcripts, the observed reduction in RNA levels may be a consequence of the effects on proliferation.

Calabretta: We have done some experiments *in vivo* in which we saw down-regulation of c-*myc* protein by c-*myc* antisense phosphorothioate oligonucleotides, but not down-regulation of a control heat-shock protein. The toxicity induced by G_4 oligonucleotides is probably cell type specific because, for example, in haemopoietic cells the effect of G_4-containing oligonucleotides is much less prominent than that on cells which are attached to substrates. This suggests that G_4 sequences might interfere with cell–extracellular matrix interactions. Of course, we did these experiments *in vitro*, so this may not be true for cells *in vivo*.

Krieg: We have looked *in vivo* at the effects of G_4 sequences on the immune system, and we have observed a prominent induction of γ-interferon (IFN-γ) by oligonucleotides containing this G_4 motif. Any phosphorothioate oligonucleotide will induce IFN-γ production but those with G_4 sequences do so at much lower doses.

Stein: I would like to state explicitly that we certainly do not know all there is to know about how G_4 sequences influence the folding of the molecules; we don't understand how folding relates to changes in protein-binding affinity; we don't understand how the sequence affects these processes; and we don't understand how all these things fit together in a cellular context. Given all that, it's difficult, if not impossible, to make any blanket statements about G_4-containing oligonucleotides. It is my firm belief that they have to be studied on an individual basis for each particular system in question.

Iversen: Has anyone replaced the guanosine residues with 7-deazaguanine, for example, which would not allow the formation of the higher-order structure, although they should still base pair? Because this may answer some of the questions about higher-order structure.

Stein: Yes, we have done such experiments. The effect depends on the oligonucleotide. In our case (with NFκB) it completely eliminated the antisense effect, demonstrating that the presumed antisense effect is really not an antisense effect at all. However, this may not be true for c-*myc*.

Eckstein: Apart from G_4 sequences, are there any other sequences that we should be aware of when designing oligonucleotides?

Stein: Bergan et al (1993) have got a sequence-specific, but not antisense, oligonucleotide that appears to inhibit the kinase activity of Bcr–Abl.

What also worries me is the increasing amount of literature on single-stranded DNA-binding proteins. I've seen about seven or eight examples over the last couple of years, and they recognize all kinds of sequence-specific motifs. Although I can't necessarily relate what goes on in the oligonucleotide area to that field specifically, what worries me is that the more oligonucleotide we deliver to the nucleus the more they may interact with these single-stranded DNA-binding proteins, and we do not know the extent to which these interactions are responsible for some of the effects that we are observing.

Nicklin: What are the current and projected costs of large-scale oligonucleotide synthesis?

Fearon: LR-3280 costs about US$350 per gram for the materials, and we envision with further improvements we should be able to reduce this to US$200–250 per gram.

Gait: If you could sell it tomorrow, how many kilograms would you require and how would you make it?

Fearon: The treatment of restenosis may require a single intramural dose of approximately 10–20 mg or less immediately after angioplasty. If there are a quarter of a million patients per year then we would need to make 2.5–5 kg/year, which is well within reach. We would probably scale-up our synthesizers to the 100 mM level (they're currently at the 10 mM level), which is feasible and would reduce our production costs.

References

Benimetskaya L, Tonkinson JL, Koziolkiewicz M et al 1995 Binding of phosphorothioate oligonucleotides to basic fibroblast growth factor, recombinant soluble CD4, laminin and fibronectin is P chirality independent. Nucleic Acids Res 23:4239–4245

Bergan R, Connell Y, Fahmy B, Neckers L 1993 Electroporation enhances c-*myc* antisense oligodeoxynucleotide efficacy. Nucleic Acids Res 21:3567–3573

Gryaznov S, Skorski T, Cucco C et al 1996 Oligonucleotide N3′→P5′ phosphoramidates as antisense agents. Nucleic Acids Res 24:1508–1514

Koziolkiewicz M, Krakowiak A, Kwinkowski M, Boczkowska M, Stec WJ 1995 Stereodifferentiation — the effect of P chirality of oligo(nucleoside phosphorothioates) on the activity of bacterial RNase H. Nucleic Acids Res 23:5000–5005

Maltese JY, Sharma HW, Vassiler L, Narayanan R 1995 Sequence context of antisense rel A/NFκB phosphorothioates determines specificity. Nucleic Acids Res 23:1146–1151

Tang JY, Roskey A, Li Y, Agrawal S 1995 Enzymatic synthesis of stereoregular (all R$_p$) oligonucleotide phosphorothioate and its properties. Nucleosides Nucleotides 14:985–990

Temsamani J, Kubert M, Agrawal S 1995 Sequence identity of the *n*–1 product of a synthetic oligonucleotide. Nucleic Acids Res 23:1841–1844

Wickstrom EL, Bacon TA, Gonzalez A, Freeman DL, Luman GH, Wickstrom E 1988 Human promyelocytic leukemia HL-60 cell proliferation and c-*myc* protein expression are inhibited by an antisense pentadecadeoxynucleotide targeted against c-*myc* mRNA. Proc Natl Acad Sci USA 85:1028–1032

Yaswen P, Stampfer M, Ghosh K, Cohen JS 1993 Effects of sequence of thioated oligonucleotides on cultured epithelial cells. Antisense Res Dev 3:67–77

Discovering antisense reagents by hybridization of RNA to oligonucleotide arrays

E. M. Southern, N. Milner and K. U. Mir

Department of Biochemistry, University of Oxford, South Parks Road, Oxford OX1 3QU, UK

Abstract. Antisense reagents have the potential to modify gene expression by interacting with DNA or mRNA to down-regulate transcription or translation. There have been a number of successful demonstrations of antisense activity *in vivo*. However, a number of problems must be solved before the method's full potential can be realized. One problem is the need for the antisense agent to form a duplex with the target molecule. We have found that most regions of mRNAs are not open to duplex formation with oligonucleotides because the bases needed for Watson–Crick base pairing are involved in intramolecular pairing. Using arrays of oligonucleotides that are complementary to extensive regions of the mRNA target, we are able to find those antisense oligonucleotides which bind optimally. There is good correspondence between the ability of an oligonucleotide to bind to its target and its activity as an antisense agent in *in vivo* and *in vitro* tests. To understand more fully the rules governing the process of duplex formation between a native RNA and complementary oligonucleotides, we have studied the interactions between tRNA^{phe} and a complete set of complementary dodecanucleotides. Only four of the set of 65 oligonucleotides interact strongly. The four corresponding regions in the tRNA share structural features. However, other regions with similar features do not form a duplex. It is clear that *ab initio* prediction of patterns of interaction require much greater knowledge of the process of duplex formation than is presently available.

1997 Oligonucleotides as therapeutic agents. Wiley, Chichester (Ciba Foundation Symposium 209) p 38–46

Antisense oligonucleotides must interact with mRNAs by Watson–Crick base pairing, stimulating RNase H cleavage or blocking translation (Stein & Cheng 1993). It is generally found that sequences targeted to different regions of a mRNA produce different levels of antisense activity. There appears to be no correlation with biological properties; the translation start site is not particularly favourable, and sites that give strong antisense response are distributed throughout the mRNA. The differences are probably due to differences in the accessibility of the RNA to heteroduplex formation as a result of intramolecular folding. Most regions of many mRNAs form quite stable intramolecular structures. In order to form heteroduplexes with such a structure, an oligonucleotide must find unpaired bases to initiate duplex formation. Propagation of the duplex requires displacement of existing base pairs in the RNA. The yield of heteroduplex is not so dependent on the difference in the free energies of the reactants and the product used in computer predictions, as it is on the detailed structure of the reactants and their potential for productive interaction.

We have used large arrays of oligonucleotides complementary to all positions in the sequence of tRNA^{phe} to explore the relationship between RNA structure and duplex yield. Arrays complementary to extensive regions of mRNAs for rabbit β-globin (Milner et al 1997) and to c-*raf* (N. Milner, unpublished experiments, 1995) were used to find oligonucleotides with different binding strengths. There was good correlation between heteroduplex yield measured on the arrays and antisense potential *in vivo* and *in vitro*.

Oligonucleotide arrays

Large numbers of oligonucleotides can be made as an array on the surface of glass or plastic sheet by a simple combinatorial protocol (Southern et al 1994) (Fig. 1). The array is used in a hybridization reaction with labelled target RNA, and the yields of heteroduplexes measured from an image obtained by exposing the array to a storage phosphor screen which is subsequently scanned in a phosphorimager. The protocol we use to make the arrays produces oligonucleotides ranging in length from mononucleotides to an upper limit that is determined by the dimension of the cell used to apply the reagents to the substrate and the displacement at each coupling step. Typically, we use arrays which 'scan' 10–150 bases of the target. For the experiments with tRNA, chosen for this study because its structure is so well characterized (Robertus et al 1974, Suddath et al 1974), we were able to make an array of all complementary monomers to 12 mers.

RNA structure and heteroduplex formation

Only four of the 65 complementary 12 mers give significant heteroduplex yield when native tRNA is hybridized to the array (Mir 1995, Mir & Southern 1997) (Fig. 2). A heteroduplex is formed with each of the four stems. Significantly, only one side of each stem forms a heteroduplex. Most of the tRNA is inaccessible to the oligonucleotides. The regions that do form a heteroduplex are characterized by having unpaired bases stacked onto the end of a stem; all or most of the bases that are incorporated into the heteroduplex are part of a continuous helix within the tRNA. Evidently this is a structure that favours initiation—at the unpaired bases—and propagation by strand displacement.

Each region of the RNA that is incorporated into a heteroduplex is included in several other overlapping oligonucleotides with which the RNA does not form a heteroduplex. These seem to be destabilized by structures in the tRNA. For example, no heteroduplex is

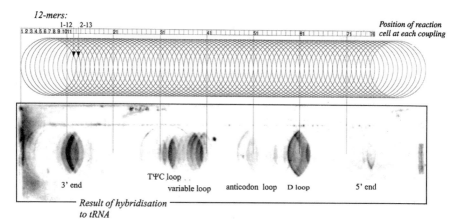

12-mers:

Result of hybridisation to tRNA

FIG. 1.   Top. An array of complementary oligonucleotides is created by coupling nucleotide residues in the order in which they occur in the complement of the target sequence using a circular reaction cell pressed against the surface of a glass plate which had been modified to allow oligonucleotide synthesis (Southern et al 1994). After each base coupling the plate is moved along by a fixed increment: in the experiment illustrated here, the array was made using 'reverse synthons', i.e. 5'-phosphoramidites, protected at the 3'-hydroxyl, leaving 5' ends of the oligonucleotides tethered to the glass. The first base was added at the left-most position. The diameter of the reaction cell was 30 mm and the offset at each step to the right was 2.5 mm. The result is that after 12 steps, an oligonucleotide complementary to bases 1–12 of the tRNA had been synthesised in a lenticular patch 2.5 mm wide. In addition, all 11 mers are in the cells flanking the 12 mers, the next row of cells contains 10 mers and so on to the edge rows which contain the 76 mononucleotides complementary to the sequence of the tRNA.

Bottom. Hybridization of tRNA^{phe}, labelled with ^{32}P prepared from native tRNA by the ligation of ^{32}P-cytidine-5'-3'-diphosphate to the 3' end. Hybridization (4 μCi total target in 1 ml solution, in 3.5 M tetramethylammonium chloride at 4 °C), detection and analysis were carried out as described in Southern et al (1994). A striking feature of the hybridization pattern is the sharp transitions in intensity: positions of strong interaction are adjacent to positions in the sequence where yield is close to background.

formed that does not displace the stem completely. Presumably, such heteroduplexes would be displaced by reformation of the stem. No heteroduplex is formed that would require the penetration of a second stem. Again, it seems likely that the formation of such heteroduplexes is inhibited by the inherent stability of the stem.

Although this study provides some insight into the mechanism of heteroduplex formation, the main lesson to be drawn from it is that the features in the RNA which control heteroduplex formation are subtle and complex. Put another way, it would not be possible to predict the hybridization pattern that we found for the tRNA from its structure. The problem in predicting the outcome for a mRNA is obvious: there is no equivalent structural information, and computer predictions of RNA folding are quite inadequate for this purpose.

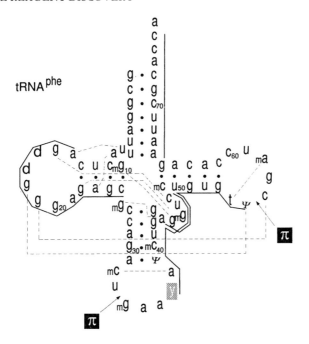

FIG. 2.   The regions of the tRNA which give highest heteroduplex yield are marked alongside the cloverleaf structure of tRNA*phe*; tertiary contacts are marked with thin broken lines. The solid lines represent the regions which give high duplex yield in tetramethylammonium chloride. Note that of the 65 potential 12 mer interactions with the target RNA, only a small number give high duplex yield. Features that would hinder duplex progression are marked.

Heteroduplex yield and antisense activity

We have analysed several RNAs, including parts of the HIV genome and several mRNAs, by hybridization to arrays and in all cases find that only a few regions are open to heteroduplex formation. The viral RNA is particularly resistant to heteroduplex formation; only one 12 mer oligonucleotide in a 76 base region of the transactivation response (TAR) formed substantial heteroduplex yield (K. U. Mir, unpublished work 1993).

We describe here experiments with rabbit β-globin, a system that has been studied as a model by others (Cazenave et al 1986), and c-*raf*—a target for antisense therapy that has been thoroughly studied by Monia et al (1996).

An array of oligonucleotides complementary to the first 122 nucleotides at the 5' end of β-globin mRNA was made on glass. The longest oligonucleotides were 17 mers. The array was hybridized with a radioactively labelled RNA made by transcription of a reverse transcriptase (RT) -PCR product obtained from purified globin mRNA (Gibco BRL Life Technologies, Paisley, UK) with a T7 promoter incorporated in one of the primers (Fig. 3). Much of this region of the RNA showed no significant hybridization. The hybridization pattern indicates that only half of the bases take part in any significant level of heteroduplex formation with their complementary oligonucleotides. The strongest interaction is with a 15 mer that includes the start codon: bases 47–61. The heteroduplex yield from this sequence, and the two 16 mers and three 17 mers that include it, was more than double the yield of the next most intense. The yield was fivefold of that given by the 17 mer used by Cazenave et al (1986) in their studies. Three oligonucleotides—that which gave the highest yield, BG1; that which was used by Cazenave, BG2; and one which gave a low yield, BG3, bases C85–U99 of the mRNA— were tested for their capacity to inhibit translation of β-globin mRNA in a wheat germ extract. BG3 showed low antisense activity. BG1 had the same activity as BG2 when used at only one-fifth to one-tenth the concentration, indicating a correlation between the hybrid yield on the array and *in vitro* antisense activity.

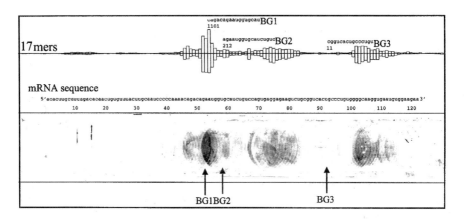

FIG. 3.   Analysis of a β-globin transcript on an array representing the first 122 bases of the mRNA. The array was fabricated on a glass plate (600 × 350 mm) using the method described in Southern et al (1994) with a circular template that produced monomers to 17 mers. RNA was transcribed from a DNA molecule made from mRNA by reverse transcriptase PCR and including a T7 promoter in the 5′ primer. α-^{33}P UTP was included in the transcription mixture. The transcript in hybridization buffer (1 M NaCl, 10 mM Tris pH 8.0, 1 mM EDTA, 0.01% SDS, 0.5 ml) was applied in a thin layer between the array and a plain glass plate of similar size. Hybridization was at 30 °C for 18 h. After rinsing in hybridization buffer, the array was exposed to a storage phosphor screen for 20 h. The screen was scanned in a Molecular Dynamics 400A PhosphorImager (Molecular Dynamics, Chesham, Bucks, UK) and the image analysed using xrseq (Wang & Elder 1997). A template comprising a series of overlapping circles is placed over the image (Fig. 1, Top) and each region within areas defined by the template is integrated. The array is symmetrical about the central axis, providing duplicate measurements for each oligonucleotide. The histogram shows the integrated pixel values for the areas on the array which carry 17 mers; the two values are for the areas above and below the centre line of the array. BG1, BG2 and BG3 are the oligonucleotides which were used for *in vitro* tests of antisense effect. Sequences represent the mRNA—the complements of the antisense oligonucleotide sequences.

Extensive comparisons of the capacity of BG1 and BG2 to stimulate RNase H activity also showed a five- to 10-fold greater activity for BG1.

Monia et al (1996) tested 34 phosphorothioate oligonucleotides for antisense activity against c-Raf, a protein in the MAP kinase cascade involved in the control of cell proliferation. c-Raf is active in a number of tumours and is thus a candidate for therapeutic intervention by antisense technology. Most of these oligonucleotides showed modest or no capacity to reduce c-*raf* mRNA levels in cells in culture. One, designated ISIS 5132, showed strong activity. Furthermore, ISIS 5132 gave potent inhibition of tumour growth when injected intravenously into nude mice that had xenografts of three different human tumours.

We analysed the hybridization behaviour of an *in vitro* transcript from a full-length copy of c-*raf* cloned in a plasmid (Fig. 4). The array of complementary oligonucleotides covered 100 bases in the 3' region of the mRNA that included ISIS 5132.

Approximately 10 oligonucleotides in this region gave appreciable heteroduplex yield: ISIS 5132 was one of two that gave the highest yield. Several of the other oligonucleotides tested by Monia et al (1996) fell within the region we analysed on the array. These all gave low heteroduplex yield and had low activity in the antisense test of Monia et al (1996).

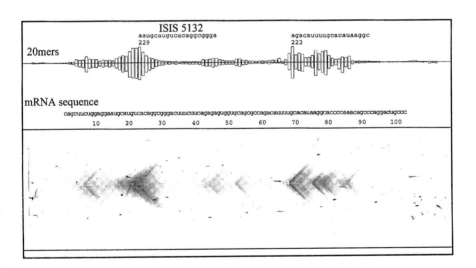

FIG. 4. A scanning array representing the complements of human c-Raf kinase mRNA was fabricated as described, on a substrate of aminated polypropylene (Southern et al 1994), with a diamond-shaped template that produced monomers to 22 mers. The transcript was made from a plasmid carrying a full-length copy of the cDNA (gift of Heinz Moser), labelled by incorporation of α-[33]P -UTP. Hybridization conditions were similar to those used for the globin transcript, but hybridization was carried out in a rotating cylinder. ISIS 5132 is one of the two 20 mers with the highest intensity.

Conclusion

We find that analysis on arrays of oligonucleotides provides a simple method of detecting oligonucleotides with potential antisense activity. In the limited study described here, there was good correlation between heteroduplex yield and antisense activity *in vitro* and *in vivo*. These results indicate that the high rate of failure of antisense reactions is most likely due to the high degree of internal base pairing in mRNAs which prevents interaction with oligonucleotides.

Acknowledgements

We thank S. C. Case-Green and J. Williams for help and advice.

References

Cazenave C, Loreau N, Toulme J-J, Hélène C 1986 Anti-messenger oligodeoxynucleotides: specific inhibition of rabbit β-globin synthesis in wheat germ extract and *Xenopus* oocytes. Biochimie 68:1063–1069

Milner N, Mir KU, Southern EM 1997 Selecting effective antisense reagents on combinatorial oligonucleotide arrays. Nat Biotechnol 15:537–541

Mir KU, Southern EM 1997 RNA structure and the mechanism of duplex formation, submitted

Mir KU 1995 Novel approaches for the analysis of nucleic acids. PhD thesis, University of Oxford, UK

Monia BP, Johnston JF, Geiger T, Muller M, Fabbro D 1996 Antitumour activity of a phosphorothioate antisense oligodeoxynucleotide targeted against c-*raf* kinase. Nat Med 2:668–675

Offensperger WB, Offensperger S, Walter E, Teubner K, Igloi G, Blum HE, Gerok W 1993 *In vivo* inhibition of duck hepatitis B virus replication and gene expression by phosphorothioate-modified antisense oligodeoxynucleotides. EMBO J 12:1257–1262

Robertus JD, Ladner JE, Finch JT et al 1974 Structure of yeast phenylalanine tRNA at 3 Å resolution. Nature 250:546–551

Southern EM, Case-Green SC, Elder JK et al 1994 Arrays of complementary oligonucleotides for analysing the hybridisation behaviour of nucleic acid. Nucleic Acids Res 22:1368–73

Stein CA, Cheng Y-C 1993 Antisense oligonucleotides as therapeutic agents—is the bullet really magical? Science 261:1004–1012

Suddath FL, Quigley GJ, McPherson A et al 1974 Three-dimensional tertiary structure of yeast phenylalanine transfer RNA. Science 185:435–439

Wang L, Elder JK 1997 Program for analysing oligonucleotide arrays, in preparation

DISCUSSION

Wagner: Were the hybridizations carried out under conditions of equilibrium?

Southern: Yes, but it's not entirely clear what the nature of the equilibrium is. There is an excess of the oligonucleotides, so the target would be in low concentration and we do the hybridizations for a long period of time.

Wagner: Do you see a change in the patterns over time?

Southern: No. There is just a build-up of the pattern during a time-course experiment.

Gewirtz: What are the physical conditions of hybridizations?

Southern: They are different in each experiment I showed you. Typically, we use 1 M NaCl and the experiment is performed at 30 °C.

Wagner: How is the tRNA folded?

Southern: We don't do anything special to fold the tRNA. We just buy it from our suppliers, so it comes already folded, and we simply take the *in vitro* transcripts and use them directly.

Gewirtz: Have you attempted to mimic the milieu inside the cell? If the folding is what dictates access, it's difficult to conceptualize the biology behind these elegant approaches because it's almost impossible to predict folding within a cell where there are a multitude of different proteins.

Southern: Unless the proteins that interact with the nucleic acids are either displaced by the oligonucleotides or are not interacting with them in the first place.

Gewirtz: But that would make the molecule fold differently yet again.

Southern: It's possible. We don't know enough yet to jump to that conclusion. If we take the tRNA itself as an example, the basic architecture of the complex is probably determined more by the RNA interactions than by the proteins, although admittedly some of the protein–RNA interactions will influence the structure. We have to start somewhere. We would like to be able to study the RNA–protein interactions, but in that case where would we start?

Gewirtz: We've tried a different approach, although admittedly it's not perfect, by doing a footprint assay in the presence of nuclear extract, which attempts to mimic what that transcript is doing to see when it's inside the nucleus. Assuming that an accessible region is capable of being digested by an enzyme, one might be able to predict which regions of that RNA are accessible in a more biological medium and then design the oligonucleotides based on these results.

Caruthers: I was surprised that the results of the c-*raf* experiment correlated with those on the native RNA transcript, which was a biologically defined piece of DNA.

Crooke: This involved an element of luck. The oligonucleotide first interacts with the RNA and then the oligonucleotide-induced RNA structure interacts with RNase H in the assay. We're learning a lot about what happens with RNase H, it is not just a simple proposition of RNase H interactions with RNA structures. Therefore, it's extremely complicated and it's surprising that there was a correlation.

Southern: Yes, there was an element of luck but out of the 100 positions we scanned there were just two that gave high duplex yields so the statistics are pretty high.

Crooke: The assay identifies the sites where there might be a positive event, from an antisense perspective. The *in vitro* experiments measuring RNA levels will then define which hits are real positives. In the Raf kinase case a correlation is interesting, but a lack of correlation is equally as interesting.

Southern: What kind of lack of correlation?

Crooke: The prediction that a site should be active but isn't is the most obvious and the most common lack of correlation. The identification of a site that's active but is not predicted is much less common. These observations should teach us a lot about the secondary and tertiary events that are controlling the process.

Cohen: I like the concept of the oligonucleotides being joined by linkers giving rise to higher specificities, and in that context I wanted to say that we recently published the synthesis of two oligonucleotides joined by a steroid bridge (Chidambaram et al 1996). One of the rationales for doing that was that most linkers are extremely flexible. Of course, if we knew the distance between the two sites we could use a more rigid linker to increase the specificity.

Hélène: How does the hybridization depend on the length of the linker between the polymer and the oligonucleotide and on the density of the oligonucleotide? Do interactions occur in some of the squares between two oligonucleotides?

Southern: Yes, they do. When we used polypropylene the oligonucleotides were extremely crowded, and we are now taking steps to put them further apart because this crowding does, to some extent, interfere with the analysis. If we use a spacer to separate them from the surface we can increase the yield 100-fold.

Hélène: Are there any changes in the pattern?

Southern: We haven't done enough experiments yet to determine the effect on the pattern.

Crooke: I have always been fascinated as to why so much nucleic acid is associated with membranes. Using your array it would be interesting to see whether there were any pattern changes as a result of such interactions.

Hélène: George Sczakiel in Heidelberg has also shown this with the HIV RNA (Homann et al 1996). They showed that it is possible to increase the hybridization rate by about 100-fold using cationic lipids such as cetyltrimethlyammonium bromide.

Southern: Some years ago we also found that tetramethylammonium chloride increased the on-rate by 30-fold compared with using sodium as a counter-ion.

Vlassov: Have you performed the tRNA hybridization experiments in the presence of magnesium?

Southern: Yes. In the presence of magnesium the two heteroduplexes that initiate in the hinge region are abolished, suggesting that the hinge region is locked rigid in the presence of magnesium. If magnesium is removed, the oligonucleotides can then get into that region. The anti-codon region becomes a little more accessible in the presence of magnesium.

References

Chidambaram N, Zhou L, Cohen JS 1996 Synthesis of steroid-linked and steroid-bridged oligodeoxynucleotides. Drug Delivery 3:27–33

Homann M, Nedbal W, Sczakiwl G 1996 Dissociation of long-chain duplex RNA can occur via strand displacement *in vitro*: biological implications. Nucleic Acids Res 24:4395–4400

Pharmacokinetics of oligonucleotides in cell culture

Bernard Lebleu, Ian Robbins, Lionel Bastide, Eric Vives and Jay E. Gee

Molecular Genetics Institute, CNRS, 1919 Route de Mende, BP 5051, F-34033 Montpellier, Cédex 1, France

Abstract. Synthetic oligonucleotides offer interesting perspectives for the regulation of gene expression in normal and pathological situations. Poor uptake in many cell types, inadequate intracellular compartmentalization, often fragmentary knowledge of intracellular behaviour and mechanism of action, and lack of specificity remain major challenges. These limitations strongly urge the design of new oligonucleotide analogues and more efficient antisense strategies. Present achievements and perspectives for further developments will be discussed with emphasis on cell delivery and intracellular fate.

1997 Oligonucleotides as therapeutic agents. Wiley, Chichester (Ciba Foundation Symposium 209) p 47–59

Synthetic oligonucleotides offer interesting and various possibilities to regulate gene expression. Almost every step in the complex series of events leading to protein synthesis can in principle be targeted knowing that oligonucleotide sequences can be engineered for the sequence-specific binding of genes, RNA transcripts or even proteins (see Crooke & Lebleu 1993 for a review). Interesting new developments include the correction of point mutations through the masking of cryptic splicing sites (Sierakowska et al 1997) or through site-specific recombination (Yoon et al 1996), and the generation of site-specific mutations during the process of triple helix repair (Wang et al 1996).

The initial enthusiasm has been somewhat tempered in recent years as the limitations and the problems inherent to these approaches have become evident to an increased number of investigators.

Major problems in the use of synthetic oligonucleotides in cell culture: an overview

Increased resistance to nuclease degradation can be achieved through various chemical modifications or through the association of unmodified oligonucleotides to delivery vectors (*vide infra*).

Increased binding to complementary nucleic acid targets has been reached through chemical modifications of the oligonucleotide backbone or through the attachment of various pendant groups. Arresting transcription or translation elongation are indeed important issues that are understandably difficult to achieve because these processes have built-in tools to unwind secondary structures efficiently. Peptide nucleic acids (PNAs) (Nielsen 1995) and N3′ phosphoramidates (Gryaznov et al 1995) provide interesting new developments in this respect.

Favouring passage through biological barriers and reaching a valuable steady-state concentration in the appropriate intracellular compartment, e.g. in the cytoplasm or in the nucleus, still represent major hurdles. These points will be discussed in more detail below.

Compromising improved pharmacological properties with other parameters required for the optimal activity of antisense oligonucleotides such as RNase H activation or target recognition with an appropriate selectivity has been difficult to achieve. Dealing with the complex and poorly understood intracellular environment is obviously more complicated than dealing with well-standardized hybridization studies in which stringency can be manipulated at will.

Binding to serum and cellular proteins (sometimes in a sequence-specific way) has been experienced with phosphorothioate derivatives in particular (Stein & Cheng 1993). Our own studies have documented protein-binding sites for charged oligonucleotides in nuclear extracts (Clarenc et al 1993).

This propensity of polyanions (and more so of sulfated polyanions) to bind to proteins is now considered to be a major problem. Oligonucleotide analogues with improved nucleic acid binding and minimal protein binding have not yet been designed to our knowledge.

New developments to try to overcome these problems include the design of chimeric oligonucleotides (Giles & Tidd 1992) and the synthesis of prodrugs as for instance t-But SATE derivatives (Mignet et al 1997), nitroveratryl derivatives (unpublished observations in our laboratory) or peptide conjugates (*vide infra*).

Another problem that has frequently been underevaluated deals with toxicity and immunogenicity of oligonucleotide analogues and of their degradation products. These considerations are major issues when designing clinical trials but have led to unexpected data *in vitro* as well. A striking example in this respect is the demonstration of a sequence-related, non-antisense effect of a *bcr–abl* oligonucleotide whose *in vitro* growth inhibitory effect was due to the release of degradation products (J. C. Vaerman & P. Martiat, personal communications 1996).

Cellular uptake and intracellular compartmentalization of synthetic oligonucleotides

Antisense oligonucleotides are taken up by various cell types and accumulate at intracellular concentrations required to regulate gene expression. Passive diffusion through a lipid bilayer does not take place at an appreciable level even for uncharged backbone-modified analogues, such as methylphosphonate derivatives, or for PNAs. The mechanisms involved in cell binding and intracellular delivery of synthetic oligonucleotides have not been studied extensively. Membrane proteins with molecular weights ranging from 75 to 110 kDa have been described by several groups as potential cell-surface receptors for synthetic oligonucleotides (Yao et al 1996). Whether this reflects the existence of a family of oligonucleotide-binding proteins or artefacts cannot be ascertained. The parts played by receptor-mediated endocytosis, by pinocytosis and possibly by mechanisms other than endocytic processes are still matters of controversy (Wu-Pong et al 1994). Likewise, the intracellular routing of synthetic oligonucleotides and the mechanisms through which they are released from endocytic compartments are not understood. Fluorescence microscopy studies of various cell lines with fluorochrome-conjugated oligonucleotides reveal a punctate cytoplasmic distribution consistent with the accumulation of oligonucleotides in endocytic vesicles. Subcellular fractionation and drugs interfering with various steps in the endocytic process have also been used. Unfortunately, each of these experimental approaches has its limitations. Studies aiming at understanding why various cell types differ tremendously in their capacity to internalize oligonucleotides or studies on the behaviour of oligonucleotides in cell mutants affected in the endocytic pathways might be valuable approaches. In this respect, the internalization of oligonucleotides is rather efficient in some cell types, as for instance in keratinocytes.

Whatever the site at which oligonucleotides are released from the endocytic compartment and the efficiency of the process, some material reaches the cytosolic compartment. The distribution of microinjected oligonucleotides between the cytoplasm and the nuclei has been monitored in various cell types with fluorescently tagged or BrdU-modified oligonucleotides. The material released in the cytoplasm rapidly diffuses through the nuclear pores and accumulates in the nuclei. Non-histone, but poorly characterized, nuclear proteins bind oligonucleotides strongly (Clarenc et al 1993). Available experimental data are consistent with the binding of oligonucleotides to components of the nuclear RNA-processing machinery and with small nuclear RNPs in particular. Nuclear accumulation could, at first sight, be considered as beneficial, in particular when targeting DNA, pre-mRNAs or DNA-binding proteins. Whether nuclear binding allows the progressive release of oligonucleotides or irreversibly traps them is essentially unknown. Protein binding by polyanionic material (and in particular by sulfated polyanions, such as phosphorothioate oligonucleotides) is a source of non-antisense effects (reviewed in Stein & Cheng 1993). Whether binding of oligonucleotides to components of the RNA-processing machinery might be detrimental for cells or might contribute to non-specific effects is unknown.

Improving oligonucleotide delivery

As mentioned above, synthetic oligonucleotides are taken up at concentrations leading to significant phenotypic responses in various experimental models. In many instances, however, poor cellular uptake and escape from the endocytic compartments severely limit or completely prevent their activity in intact cells. Various strategies have been used in order to overcome these problems.

Simple chemical modifications of the oligonucleotide backbone, which appreciably improved other pharmacological properties of oligonucleotides, have not been successful. As an example, neutral backbone oligonucleotides, such as methylphosphonates, are not taken up more readily than charged oligonucleotides, and they suffer from poor solubility in an aqueous environment.

In most cases existing drug delivery vehicles have been adapted to the specific requirements of oligonucleotide administration. Increasing interest in the development of non-viral gene delivery vectors (which bear similar objectives and problems) has given a strong impetus to this field in recent years. The chemical conjugation or the physical association of oligonucleotides to the carrier protects them from nuclease degradation. This eventually allowed the use of unmodified oligonucleotides, which might be beneficial knowing the problems (e.g. non-specific binding to proteins) or the limitations (e.g. incapacity to activate RNase H) encountered with many oligonucleotide analogues.

Previous work in our group had established the potential of polycationic polymers, such as poly (L-lysine), for the cytoplasmic delivery of biologically active 2′,5′-linked adenylic oligomers. Phosphodiester oligonucleotides (15–17 mer in most experiments) chemically conjugated to poly (L-lysine) were more efficient (50–100-fold in most cases) than non-conjugated material in several biological models (see Lebleu et al 1997 for a review). The increased efficiency essentially arose from increased membrane association and subsequent internalization, and to some extent to increased nuclease resistance. Interestingly, oligonucleotides targeted to viral mRNAs or to the viral genomic RNA exerted a sequence-specific antiviral activity in HIV1-infected T lymphocytes, whereas non-vectorized oligonucleotides had little specificity. Poly (L-lysine)-based delivery vectors eventually allow cell targeting when conjugated to proteins as transferrin or asialoglycoproteins (Wu & Wu 1992). Cytotoxicity in some cell lines and, more importantly, complement activation will unfortunately limit the use of poly (L-lysine)-based delivery vehicles for the systemic delivery of nucleic acids. Interestingly, the co-administration of sulfated polyanions (as heparins) increases antiviral activity and decreases cytotoxicity in HIV-infected cells (Degols et al 1994).

Cell targeting can also be achieved through encapsidation in antibody-conjugated liposomes. We have demonstrated the usefulness of this liposome formulation for the delivery of nucleic acid material such as 2′,5′-linked adenylic oligomers, $(rI)_n.(rC)_n$ (a double-stranded RNA synthetic made by the hybridization of poly I to poly C) or antisense oligonucleotides. Incubation of the target cells with the appropriate antibodies was strictly required, and full protection of the encapsidated material against nucleases was demonstrated in several *in vitro* models (reviewed in Clarenc et al 1993). Unmodified antisense oligonucleotides delivered as a liposome formulation exerted a sequence-specific antiviral activity in *de novo* HIV-infected cells, whereas phosphorothioate oligonucleotide derivatives inhibited virus replication through different mechanisms (Zelphati et al 1993). Here again, appropriate delivery has led to sequence-specific inhibition of gene expression at submicromolar concentrations even when using unmodified oligonucleotides. However, conjugation to cationic polymers, such as poly (L-lysine), or association with particulate carriers, such as antibody-targeted liposomes, does not bypass a major barrier in oligonucleotide delivery, e.g. efficient escape from the endocytic compartment. This is also a critical issue for gene delivery with non-viral vectors.

We have attempted to circumvent this problem with other liposome formulations. As an example, pH-sensitive liposomes should release the entrapped material through destabilization of the endosomal membrane or through membrane fusion. Antiviral activity in vesicular stomatitis virus-infected cells was obtained in a concentration range at which free oligonucleotides were not active (Milhaud et al 1996). Lipid toxicity, stability of pH-sensitive liposomes in the presence of serum, efficiency of encapsidation and preferential uptake of particulate material by macrophages are the main limitations.

Much effort is now being done to mimic viruses that have evolved efficient strategies for the delivery of their genome into cells. A well-documented case is the influenza virus, whose envelope-associated haemagglutinin undergoes a series of pH-triggered conformational changes, leading to the exposure of a fusogenic N-terminal domain in the endosomes. Although detailed mechanisms are not understood, this leads to the fusion and/or destabilization of the endosomal membrane and the release of viral genetic material in the cell cytoplasm (see White 1992 for a review). The endosome-destabilizing properties of the influenza peptide have been used to improve the efficiency of plasmid DNA delivery by transferrin–poly (L-lysine) conjugates (Planck et al 1994). A similar strategy has been used to improve the delivery of antisense oligonucleotides. An antisense oligonucleotide modified at its 5′ end by a hexamethylene-linked pyridyldisulfide group was chemically conjugated to the N-terminal fusogenic domain of the influenza haemagglutinin modified with an additional C-terminal cysteine residue. The opening of the disulfide bridge in the reducing intracellular milieu should dissociate the nucleopeptide and release the oligonucleotide in the cell cytoplasm. The pH-dependent disruption of a lipid bilayer and an increased antiviral activity in the HIV model described above have been demonstrated (Bongartz et al 1994). Combining fusogenic peptides with cationic delivery vehicles (or with particulate systems) might be worth exploring to optimize antisense oligonucleotide or ribozyme delivery (similar strategies are being explored to improve the efficiency of plasmid DNA, i.e. gene, delivery). Mimicking the topology of a viral particle might, however, be difficult to achieve.

Exploiting the cellular membrane-crossing ability of some viral or cellular proteins might represent an interesting alternative. As an example, the homeodomain of the *Drosophila* protein Antennapedia is internalized by neurones, and it accumulates in cell nuclei through an energy-independent mechanism. These translocation properties have been assigned to a 16 amino acid peptide corresponding to the third helix of the homeodomain (Derossi et al 1996). Remarkably this peptide is able to improve the delivery of conjugated peptides or antisense oligonucleotides.

Likewise, the HIV TAT transactivation protein is taken up by cells. This translocation activity has been ascribed to a region known as the TAT protein basic domain. A peptide extending from residue 37 to 72 allowed the internalization of conjugated proteins (Fawell et al 1994). Plasma membrane translocation and nuclear accumulation have been assigned to a small cluster of basic amino acids containing a nuclear localization signal (Vives & Lebleu 1997, Vives et al 1997). Whether such peptides might be helpful for the intracellular delivery of antisense oligonucleotides or of PNAs is presently being evaluated in our group. Such basic peptides might also be interesting tools to improve the hybridization properties of antisense oligonucleotides and PNAs.

Chemical conjugation to lipophilic groups and in particular to cholesterol moieties has been described by several groups. Increased resistance to nucleases and increased uptake, in part through association with seric low density lipoprotein particles, are responsible for improved biological activity (Krieg et al 1993).

Physical association to various cationic lipids formulations (see Behr 1994 for a review) appears to be easily implemented, they provide efficient strategies for plasmid DNA and antisense oligonucleotide delivery, at least in cell culture experiments. These cationic lipids act as 'double-sided sticky tape', bridging the transported nucleic acid material and the phospholipid bilayer of the plasma membrane. Titration of endosomal protons by amine groups in these carriers probably leads to endosome destabilization and cytoplasmic release of the transported nucleic acids. In keeping with this hypothesis, polyethyleneimine turned out to be as efficient as cationic lipids (Boussif et al 1995). Interesting new developments include the association of targeting moieties (J. P. Behr, personal communication 1996) or new formulations eliciting improved delivery in the presence of serum (Lewis et al 1996).

Conclusion

The intracellular behaviour and the mechanism of action of synthetic oligonucleotides are still far from being understood, despite numerous studies. Crucial problems, such as non-specific protein binding or delivery to the appropriate intracellular compartment, have not found adequate solutions, at least in the perspective of *in vivo* use. An increased knowledge of cellular pharmacology will hopefully allow further progress in the design of more efficient and more specific oligonucleotides, and eventually in the design of suitable delivery vectors.

Acknowledgements

This work was financed by the Centre National de la Recherche Scientifique, and by grants from the Agence Nationale de Recherche sur le SIDA and the Association pour le Recherche contre le Cancer.

References

Behr JP 1994 Gene transfer with synthetic cationic amphiphiles: prospects for gene therapy. Bioconjugate Chem 5:382–389

Bongartz JP, Aubertin AM, Milhaud PG, Lebleu B 1994 Improved biological activity of antisense oligonucleotides conjugated to fusogenic peptides. Nucleic Acids Res 22:4681–4688

Boussif O, Lezouala'h F, Zanta MA et al 1995 A versatile vector for gene and oligonucleotide transfer into cells in culture and in vivo: polyethyleneimine. Proc Natl Acad Sci USA 92:7297–7301

Clarenc JP, Degols G, Leonetti JP, Milhaud P, Lebleu B 1993 Delivery of antisense oligonucleotides by poly (L-lysine) conjugation and liposome encapsulation. Anticancer Drug Des 8:81–94

Crooke ST, Lebleu B 1993 Antisense research and applications. CRC Press, Boca Raton, FL

Degols G, Devaux C, Lebleu B 1994 Oligonucleotide poly (L-lysine) heparin complexes: potent sequence-specific inhibitors of HIV-1 infection. Bioconjugate Chem 5:8–13

Derossi D, Calvet S, Trembleau A, Brunissen A, Chassaing G, Prochiantz A 1996 Cell internalization of the third helix of the Antennapedia homeodomain is receptor-independent. J Biol Chem 271:18188–18193

Fawell S, Seery J, Daikh Y et al 1994 TAT-mediated delivery of heterologous proteins into cells. Proc Natl Acad Sci USA 91:664–668

Giles RV, Tidd DM 1992 Increased specificity for antisense oligodeoxynucleotide targeting of RNA cleavage by RNase H using chimeric methylphosphonodiester/phosphodiester structures. Nucleic Acids Res 20:763–770

Gryaznov SM, Lloyd DH, Chen J-K et al 1995 Oligonucleotide N3'→P5' phosphoramidates. Proc Natl Acad Sci USA 92:5798–5802

Krieg AM, Tonkinson J, Matson S et al 1993 Modification of antisense phosphodiester oligodeoxynucleotides by a 5'-cholesteryl moiety increases cellular association and improves efficacy. Proc Natl Acad Sci USA 90:1048–1052

Lebleu B, Bastide L, Bisbal C et al 1997 Poly (L-lysine)-mediated delivery of nucleic acids. In: Gregoriadis G, McCormack B (eds) Targeting of drugs: strategies for oligonucleotide and gene delivery in therapy. Plenum Press, New York, p 115–122

Lewis JG, Lin K-Y, Kothavale A et al 1996 A serum-resistant cytofectin for cellular delivery of antisense of oligodeoxynucleotides and plasmid DNA. Proc Natl Acad Sci USA 93:3176–3181

Mignet N, Tosquelas G, Barber I, Morvan F, Rayner B, Imbach JL 1997 The pro-oligonucleotide approach: synthesis and stability of chimeric pro-oligonucleotides in culture medium and in total cell extract. New Engl J Chem 21:73–79

Milhaud PG, Bongartz JP, Lebleu B, Philippot JR 1996 pH-sensitive liposomes and antisense oligonucleotide delivery. Drug Deliv 3:67–73

Nielsen PE 1995 DNA analogs with nonphosphodiester backbones. Annu Rev Biophys Biomol Struct 24:167–183

Planck C, Oberhauser B, Mechtler K, Koch C, Wagner E 1994 The influence of endosome disruptive peptides on gene transfer with synthetic virus like gene transfer systems. J Biol Chem 269:12918–12925

Sierakowska H, Sambade MJ, Agrawal S, Kole R 1997 Repair of thalassemic human beta globin mRNA in mammalian cells by antisense oligonucleotides. Proc Natl Acad Sci USA, in press

Stein CA, Cheng Y-C 1993 Antisense oligonucleotides as therapeutic agents— is the bullet really magical? Science 261:1004–1012

Vives E, Lebleu B 1997 Selective coupling of a highly basic peptide to an oligonucleotide. Tetrahedron Lett 38:1183–1186

Vives E, Brodin P, Lebleu B 1997 A truncated HIV-1 Tat protein basic domain rapidly translocates through the plasma membrane and accumulates in the cell nucleus. J Biol Chem 272:16010–16017

Wang G, Seidman MM, Glazer PM 1996 Mutagenesis in mammalian cells induced by triple-helix formation and transcription-coupled repair. Science 271:802–805

White JM 1992 Membrane fusion. Science 258:917–924

Wu GY, Wu CH 1992 Specific inhibition of hepatitis B viral gene expression *in vitro* by targeted antisense oligonucleotides. J Biol Chem 267:12436–12439

Wu-Pong S, Weiss JL, Hunt CA 1994 Antisense c-*myc* oligonucleotide cellular uptake and activity. Antisense Res Dev 4:155–163

Yao GQ, Corrias S, Cheng YC 1996 Identification of two oligodeoxyribonucleotide-binding proteins on plasma membranes of human cell lines. Biochem Pharmacol 51:431–436

Yoon K, Cole-Strauss A, Kmiec EB 1996 Targeted gene correction of episomal DNA in mammalian cells mediated by a chimeric RNA-center-dot-DNA oligonucleotide. Proc Natl Acad Sci USA 93:2071–2076

Zelphati O, Zon G, Leserman L 1993 Inhibition of HIV-1 replication in cultured cells with antisense oligonucleotides encapsulated in immunoliposomes. Antisense Res Dev 3:323–338

DISCUSSION

Gewirtz: It is possible that single peptides will not be effective because fusogenic engines need to be tripeptides or tetrapeptides. Did you encounter such problems with the influenza peptide?

Lebleu: We were not satisfied with our initial experiments with the influenza fusogenic peptide. We observed some improvement but it was far from satisfactory, probably because a precise arrangement, or at least a sufficient local density of fusogenic peptides, is necessary.

Stein: The TAT fusogenic peptide also has a basic C-terminal end, which will interact strongly with the oligonucleotide and will change the nature of how that complex molecule interacts with cellular constituents.

Lebleu: Two strategies can be envisaged. If a disulfide linker is used, as planned initially, the two entities should, in principle, separate in the intracellular environment. Alternatively, stable bonds can be introduced between the carrier peptide and the oligonucleotide.

Stein: I'm not so sure that this would be the case because a basic peptide and an acidic oligonucleotide will interact strongly.

Lebleu: It is possible that these basic amino acid carriers are more appropriate for poorly charged oligonucleotides or for peptide nuclear acids.

Rossi: I am intrigued by the observation that oligonucleotides diffuse into the nucleus. Is there evidence to support that this is a diffusion process, as opposed to an active process?

Lebleu: I wouldn't be surprised if diffusion alone were responsible because it is known that some small proteins can pass freely through nuclear pores. We have also performed a series of experiments which showed that it is a non-energy-dependent process.

Rossi: This suggests that nucleic acids preferentially reside in the nucleus, and that there has to be some active mechanism to keep them out.

Lebleu: We don't know if this is the case, but a process that maintains oligonucleotides outside the nucleus would be beneficial.

Gewirtz: Are there any data which suggest that oligomers are more effective in keratinocytes than in other cell types?

Crooke: Bennett and colleagues have reported that active intercellular adhesion molecule (ICAM) inhibition occurs in keratinocytes in the absence of cationic lipids (see Crooke & Bennett 1996).

Wagner: Dan Chin published a paper recently on the differentiation state of keratinocytes. They showed that only those keratinocytes which expressed apoptotic markers were able to take up oligonucleotides into the nucleus. Since endosome destabilization takes place during cell death, nuclear uptake by oligonucleotides is likely an indication that the cell is differentiating and dying (Giachetti & Chin 1996).

Crooke: This work has been done both *in vitro* and *in vivo*, and it suggests that there is a significant uptake of oligonucleotides in keratinocytes, even if they are not expressing apoptotic markers.

Gewirtz: Bearing in mind that oligonucleotides will bind nuclear proteins, do these papers demonstrate conclusively that an antisense effect is occurring?

Crooke: We are not reporting that the data suggest an antisense mechanism is occurring. Diffusion may be occurring, but in order to accept that hypothesis we need to show that there are free oligonucleotides in the cell, and I don't know of any data which support that. My current view is that oligonucleotides travel up an affinity gradient from one protein to the next towards their ultimate target, which is the target that we design them for. We have clearly shown that they can associate simultaneously with multiple proteins, and the molecules that are most likely to have the highest affinity for oligonucleotides, next to RNA, are RNA-binding proteins because oligonucleotides like clustered positive charges.

Matteucci: What experiments would ascertain that?

Crooke: An interesting approach, rather than trying to prove that there's no free oligonucleotide, is to examine whether they are capable of binding reversibly and simultaneously with multiple proteins, and then to demonstrate that migration occurs from one low affinity site to a higher affinity site over time, both in cell-free systems and in cells. We have done these experiments. We have looked at oligonucleotides bound to albumin, for example, and showed that this binding can be displaced by binding to thrombin, which has higher affinity, and that this binding can be displaced by binding to zinc finger proteins.

Lebleu: But in an intact cell the affinity is higher than that which has been described for albumin.

Stein: Exactly. Albumin is a rather weak oligonucleotide-binding protein. The highest affinities occur for heparin-binding proteins. There's also evidence which suggests that the off-rates of phosphorothioates bound to heparin-binding proteins are extremely slow; therefore, the affinity gradient process you propose may well be occurring but one cannot necessarily assume that it must occur.

Crooke: The point I made is only that there are low affinity sites and higher affinity sites. An answer is obtained by looking at off-rates in the absence of competing molecules but it is crucial to look at off-rates in the context of competing binding units. Another experiment would be to look at pre-bound oligonucleotides and ask what is the off-rate from a pre-bound high or low affinity site in the presence of other binding sites.

Wickstrom: Spin-label experiments may be able to shed some light on this. A spin-labelled oligonucleotide that was in solution and was not complexed would give a high mobility signal, whereas one in the protein-bound pool would give a broader signal. If free oligonucleotides were not detected then this would support Stanley Crooke's observations.

Lebleu: There might be some particular stages at which there are free oligonucleotides. For instance, in the transit between oligonucleotides being released from endosomal compartments and their entry into the nucleus there are only a few cytoplasmic proteins that can bind to oligonucleotides.

Stein: It depends on the concentration of the proteins and the binding affinities. If the concentration is low there may be free oligonucleotides, unless cationic lipids are present.

Hélène: There are about 10^6 nuclear proteins which bind oligonucleotides with a K_d of 10^{-10} M. This represents a protein concentration of about 1 μM, suggesting that there is a sink for oligonucleotides in the nucleus.

Iversen: But even though there are similar transcript numbers for a large number of the antisense targets, it is interesting that for phosphorothioates the efficacy of phosphorothioates is generally within a small dynamic range, i.e. about 0.5–3 μM. This may be true for most people's experimental culture systems, or they may also be looking at 50 transcripts per cell or even 2000, which is a much larger concentration range and suggests that this sink is probably responsible for the bulk of the oligonucleotides in the cell. The active portion is probably too difficult to observe in the dynamic range.

Wickstrom: But that's the reservoir rather than the sink.

Rossi: Do DNA methylphosphonate analogues also concentrate in the nucleus?

Lebleu: The situation for methylphosphonates is more complicated in that they are more evenly distributed between the nuclei and cytoplasm in such microinjection experiments.

Caruthers: Do they still contain phosphate residues, i.e. are they still negatively charged?

Lebleu: Yes. They usually carry one terminal phosphodiester linkage for solubility.

Rossi: It is possible, because of the diffusion of charged nucleic acids into the nucleus, that mechanisms other than RNase H are utilized. In this case we should focus on mechanisms that might block splicing, transport or other nuclear-specific processes. Blocking translation is probably not practical.

Lebleu: That is a controversial issue. There is little evidence to suggest that oligonucleotides directed towards the coding region of an mRNA interfere directly with translation. RNase H may cleave the pre-mRNA, thereby interfering with the transport process. We cannot distinguish between these possibilities.

Crooke: We used 2'-modified oligonucleotides that don't support RNase H to demonstrate a clear reduction in protein levels at selected sites (Bennett et al 1994). It's difficult to argue that translation inhibition is not the mechanism.

Akhtar: Is there any evidence for the nuclear export of oligonucleotides?

Iversen: Yes, we have observed histone proteins in the urine of patients who have been treated with oligonucleotides.

Wagner: We tried microinjecting oligonucleotides of different lengths into cells, and we saw that after microinjection the oligonucleotides entered the nucleus but that over time they were exocytosed out of the cell. This process occurred more rapidly for shorter oligonucleotides.

Crooke: Paul Nicklin and myself are currently writing a paper showing that if proteins that bind to oligonucleotides are added extracellularly after the oligonucleotide has entered the cell then the efflux of the oligonucleotides occurs more rapidly.

Cohen: I have some unpublished results that are relevant to this discussion. Oligonucleotide delivery is clearly a major problem because the number of molecules that actually reach the target is small. Several years ago Frank Bennett published a paper on the complexation of oligonucleotides with cationic lipids, which resulted in their widespread use (Bennett et al 1992). I set up a collaboration with Yechezkel Barenholz and his group at the Hadassah Medical School, who have a system in which they use a fluorescently tagged cationic lipid that acts as a reporter molecule in the liposome, and they have looked at some of the oligonucleotides that we have prepared (Clerq et al 1997). They measure the change in fluorescence, which reflects the ratio of the oligonucleotide to the charged lipid, and they also look at the turbidity, which indicates whether there is any change in the liposome itself, i.e. whether it's aggregating or falling apart. The determining factor is the molar ratio of charge, so that if you're below the charge density of the cationic lipid, i.e. when you don't have neutralization of the cationic lipid by the oligonucleotide, you observe fluorescence quenching. This is due to the formation of channels in the surface, whereby water goes down the channels and affects the fluorescence of the reporter molecule. Thus, at a lower than 1 : 1 ratio of charge, domains of oligonucleotide and cationic groups form, there is fluorescence quenching and the liposomes break down; whereas above a 1 : 1 ratio of charge domains do not form, the surface binding becomes uniform and the liposomes remain stable. Therefore, this sheds some light not only on what is happening to the liposomes, but also on the ratio of maximum oligonucleotide delivery in these systems.

Nicklin: In my opinion we should concentrate our delivery efforts where there are problems rather than where there aren't problems. We have compelling evidence that cellular uptake occurs *in vivo*; therefore, the time spent looking at delivery from a cellular uptake point of view could be misplaced. There are issues, such as acute side-effects and inappropriate distribution, that have higher priorities.

Caruthers: I'm a cell biologist and I want to use antisense techniques as tools to understand how the cell functions, as well as therapeutic drugs.

Nicklin: There are well-documented methods for this already.

Caruthers: Yes, and electroporation is one of these, but for cell biologists to use this technology to do basic research much work has to be done on understanding how DNA enters cells.

Lebleu: There are data on cellular uptake *in vitro* — for example, in keratinocytes — but I am doubtful that the issues of cellular uptake *in vivo* are far from being resolved.

Stein: I'm interested in studying prostate cancer. However, prostate cells *in vitro* are difficult to work with. There are three prostate cell lines, which take up and sequester oligonucleotides but I've yet to see a paper in the literature demonstrating efficacy with any oligonucleotide.

Nicklin: We have shown that it is possible *in vivo*.

Stein: But you can't do it *in vitro*.

Iversen: We have also found a problem with cell culture which might be appropriate to discuss here. In our oligonucleotide experiments we did not see as robust a response in our patients as we saw in cell culture. Therefore, we lowered the oxygen concentration in our cell cultures because ambient air contains 20% oxygen and, according to Le Chatelier's principle, the amount of radical oxygen that this produces is much more than is ever present *in vivo*. This has an important impact on the adducts that are formed on membranes (lipid peroxides), and the way the cells handle a lot of products is dependent on whether they can maintain a chemically reducing environment.

Cohen: This is not the first time that *in vitro* studies result in artefacts that are not necessarily relevant to the *in vivo* situation.

Lebleu: It is therefore vital to understand the mechanisms by which oligonucleotides are picked up a cell, rather than just to accept that they are; and the study of keratinocytes is a good starting point. There has to be an active mechanism. It cannot occur by diffusion alone.

Crooke: I don't agree. It's conceivable, through shuttling from one protein to another, that a process which requires live cells, functional proteins at varying concentrations and membrane fluidity can take place that is not, in the classic sense, an active transport process, i.e. not driven by ATP hydrolysis. We should study both *in vitro* and *in vivo* systems but, in my opinion, our fundamental challenge *in vitro*, both from a cell biology point of view and a drug screening point of view, is how to find efficient means of getting non-phosphorothioates into cells. Much of our effort at the moment is directed towards identifying the cationic lipid analogue for RNA.

Nicklin: In terms of oligonucleotide uptake mechanisms, we've clearly shown *in vivo* in the liver that scavenger receptors are involved (A. Steward & P. L. Nicklin, unpublished work 1996).

Crooke: We're also trying to look at sub-organ distribution in a way that will allow us to begin to address mechanisms. It's early days yet but we're finding some interesting results in the liver.

References

Bennett CF, Chiang MY, Chan H, Shoemaker JE, Mirabelli CK 1992 Cationic lipids enhance cellular uptake and activity of phosphorothioate antisense oligonucleotides. Mol Pharmacol 41:1023–1033

Bennett CF, Condon TP, Grimm S, Chan H, Chiang MY 1994 Inhibition of endothelial cell adhesion molecule expression with antisense oligonucleotides. J Immunol 152:3530–3540

Clerq S, Lerner D, Cohen JS, Barenholz Y 1997 Insight into the interactions of oligonucleotides with liposomes. Oligonucleotide delivery, controlled release society conference. Stockholm, June 1997 (abstr)

Crooke ST, Bennett CF 1996 Progress in antisense oligonucleotide therapeutic. Annu Rev Pharmacol Toxicol 36:107–129

Giachetti C, Chin DJ 1996 Increased oligonucleotide permeability in keratinocytes of artificial skin correlates with differentiation and altered membrane function. J Investig Dermatol 106:1–7

Pharmacokinetics of oligonucleotides

Sudhir Agrawal and Ruiwen Zhang*

*Hybridon Inc., 620 Memorial Drive, Cambridge, MA 02139, and *University of Alabama at Birmingham, Department of Pharmacology and Toxicology, UAB Station, Box 600, Volker Hall 101, Birmingham, AL 35294, USA*

Abstract. The effectiveness of antisense oligonucleotides as therapeutic agents depends on their pharmacokinetics, tissue disposition, stability, elimination and safety profile. Pharmacokinetic data allow one to determine the frequency of administration and any potential toxicity associated with chronic administration. Phosphorothioate oligonucleotides degrade from the 3' end, the 5' end, and both the 3' and 5' ends in a time- and tissue-dependent manner. After intravenous administration in mice, rats and monkeys, phosphorothioate oligonucleotides are detected in plasma; they distribute rapidly and are retained in the majority of tissues. The major route of elimination is the urine. The pharmacokinetic profile is similar following subcutaneous, intradermal or intraperitoneal administration, but with lower maximum plasma concentrations. Phosphorothioate oligonucleotides have a short plasma half-life in humans. End-modified, mixed-backbone oligonucleotides (MBOs) contain nuclease-resistant 2'-O-alkylribonucleotides or methylphosphonate internucleotide linkages at both the 3' and 5' ends of phosphorothioate oligonucleotides. These end-modified MBOs have pharmacokinetic profiles similar to those of the parent phosphorothioate oligonucleotides, but they are significantly more stable *in vivo* and they can be administered orally. Centrally modified MBOs contain modified RNA or DNA in the centre of a phosphorothioate oligonucleotide. They show controlled degradation and elimination following administration in rats. The pharmacokinetics of antisense oligonucleotides depends on the sequence, the nature of the oligonucleotide linkages and the secondary structure.

1997 Oligonucleotides as therapeutic agents. Wiley, Chichester (Ciba Foundation Symposium 209) p 60–78

Oligonucleotides and their various analogues are being studied as novel therapeutic agents based on their ability to hybridize to complementary DNA or RNA and interfere with the processing of the nucleic acid, including transcription, splicing and translation (Zamecnik 1996). For an oligonucleotide to be an effective therapeutic agent, it must be bioavailable *in vivo*. Studies using phosphodiester oligonucleotides show promising results as therapeutic agents *in vitro*; however, their rapid degradation *in vivo* limits their use (Goodchild et al 1991, Sands et al 1994, Agrawal et al 1995a). A number of oligonucleotide analogues that are more nuclease resistant have been synthesized and studied for their effectiveness as antisense agents.

Phosphorothioate oligonucleotides, which have a sulfur substituted for one of the non-bridged oxygens of the internucleotide linkage, have shown promising results as antisense agents *in vitro* and *in vivo* (reviewed in Agrawal 1996). Experience gained with phosphorothioate oligonucleotides has allowed various mixed-backbone oligonucleotides (MBOs) to be designed, with improved properties as antisense agents (Agrawal et al 1990, Monia et al 1993, Metelev et al 1994, Agrawal et al 1997a).

Pharmacokinetics is a quantitative study that describes the fate of a drug *in vivo*, including absorption, distribution, metabolism and elimination (Fig. 1). Understanding the pharmacokinetics of an oligonucleotide provides the basis for its effective use as a therapeutic agent, as well as indicating which organs may be sites of toxicity. In this chapter we discuss the pharmacokinetic studies of phosphorothioate oligonucleotides and MBOs carried out in our laboratory (Fig. 2). Details about pharmacokinetic studies of oligodeoxynucleotide methylphosphonates (Chem et al 1990) and other phosphorothioate oligonucleotides or analogues can be obtained elsewhere (Sands et al 1995, Iversen et al 1994, Cossum et al 1993, Crooke et al 1996, Akhtar & Agrawal 1997).

Methodology

The choice of radiolabel and its site of incorporation in an oligonucleotide is important for oligonucleotide pharmacokinetic studies. ^{14}C-labelled oligonucleotides (Cossum et al 1993) have shown significantly different pharmacokinetic profiles from ^{35}S-labelled (Agrawal et al 1991, 1995a, Agrawal 1996, Iversen et al 1994) or ^{3}H-labelled (Sands et al 1994) oligonucleotides. In our studies we incorporated ^{35}S into oligonucleotides and

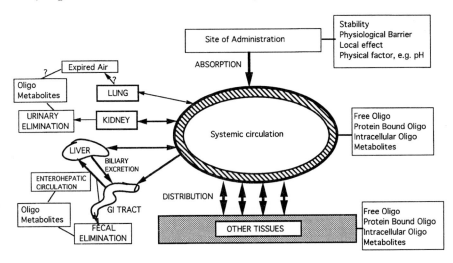

FIG. 1. Diagrammatic presentation of absorption, distribution, metabolism and elimination of oligonucleotides. GI, gastrointestinal.

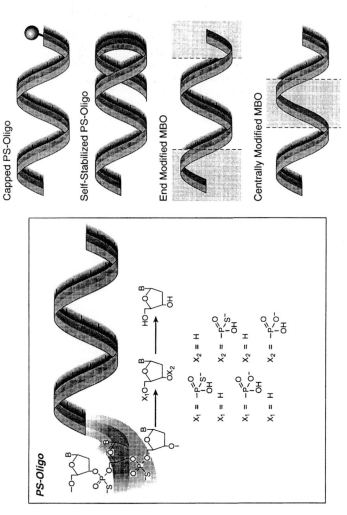

Capped PS-Oligo

Self-Stabilized PS-Oligo

End Modified MBO

Centrally Modified MBO

PS-Oligo

$X_1 = -\overset{O}{\underset{OH}{\overset{\|}{P}}}-S^-$ $X_2 = H$

$X_1 = H$ $X_2 = -\overset{O}{\underset{OH}{\overset{\|}{P}}}-S^-$

$X_1 = -\overset{O}{\underset{OH}{\overset{\|}{P}}}-O^-$ $X_2 = H$

$X_1 = H$ $X_2 = -\overset{O}{\underset{OH}{\overset{\|}{P}}}-O^-$

FIG. 2. Diagrammatic representation of a phosphorothioate oligonucleotide (PS-Oligo) and its modified analogues. The chemical structure has three distinct domains: internucleotide linkage, sugar and bases. Phosphorothioate internucleotide linkages provide a polyanionic nature and are also the site of degradation from the 3′ end, the 5′ end, or both the 3′ and 5′ ends of the oligonucleotide. Degradation products include shorter lengths of phosphorothioate oligonucleotides (not shown) as well as nucleotides and nucleosides. Various modifications of phosphorothioate oligonucleotides that protect against nuclease attack have been studied, including 3′ capped phosphorothioate oligonucleotides (carrying a modification at the 3′ end), self-stabilized oligonucleotides (carrying a hairpin loop at the 3′ end), end-modified, mixed-backbone oligonucleotides (MBOs) (carrying segments of RNA/DNA at the 3′ and 5′ ends) or centrally modified MBOs (carrying a segment of modified RNA or DNA in the centre).

their analogues, either throughout the molecule or only at internal internucleotide linkages (Agrawal et al 1996).

Animal models

We administered ^{35}S-labelled oligonucleotides to mice or rats in physiological saline intravenously (i.v.), intraperitonally (i.p.), subcutaneously (s.c.) or orally (p.o.). At various time points after administration we collected blood samples in heparinized tubes and measured the radioactivity levels. At designated time points we sacrificed the animals and collected various tissues, which we then homogenized prior to measuring the radioactivity levels. Measurement of radioactivity provided the concentration of radioactivity expressed as oligonucleotide equivalents in plasma and various tissues, but did not provide an indication of whether the radiolabel was associated with the intact or degraded form of the oligonucleotide, or not associated at all. We extracted the oligonucleotides from plasma and various tissues by phenol extraction, and analysed them using various methods, including high performance liquid chromatography (HPLC) and polyacrylamide gel electrophoresis (PAGE). Analysis of the extracted oligonucleotide by these techniques (HPLC, PAGE or capillary gel electrophoresis) provides only a relative profile of integrity of radioactivity, as extraction efficiency from various tissues is not consistent or quantitative at various post-administration time points.

Pharmacokinetics of phosphorothioate oligonucleotides

Plasma clearance, tissue disposition and elimination

Pharmacokinetics of the phosphorothioate oligonucleotide 1 (Fig. 3) following i.v. administration in mice showed rapid elimination from the plasma compartment (Agrawal et al 1991). This phosphorothioate oligonucleotide was distributed to the majority of organs, but in certain tissues, e.g. kidney, liver, bone marrow and spleen, higher concentrations were observed and retained for an extended period. The major route of elimination of Oligonucleotide 1 following i.v. or i.p. administration was in the urine. A pharmacokinetic profile similar to the one observed with i.v. administration was observed with i.p. administration, except that the peak plasma concentration was lower. Analysis of the extracted phosphorothioate oligonucleotide showed the presence of both intact and degraded forms. The degradation profile was plasma, tissue and time dependent. Higher molecular weight degradation products of the administered oligonucleotide were present in the liver and the gastrointestinal tract (Agrawal et al 1991, 1995a).

Pharmacokinetic analysis of the phosphorothioate oligonucleotide Oligonucleotide 2 (Fig. 3) in rats following i.v. administration showed a similar profile to that observed in mice (Zhang et al 1995a). This phosphorothioate oligonucleotide was rapidly cleared from the plasma compartment and was distributed to the majority of tissues.

Oligonucleotide number	**Sequence**	**Length**	**Route**	**Species**
1	5'ACACCCAATTCTGAAAATGG 3'	20-mer	IV, IP	mice
2	CTCTCGCACCCATCTCTCTCCTTCT	25-mer	IV, SC / IV	rats / monkeys, humans
3	**X**-ACACCCAATTCTGAAAATGG	20-mer	IV	mice
4	ACACCCAATTCTGAAAATGG-**X**	20-mer	IV	mice
5	**X**-ACACCCAATTCTGAAAATGG-**X**	20-mer	IV	mice
6	ACACCCAATTCTGAAAATGG-**Y**	20-mer	IV	mice
7	ACACCCAATTCTGAAAATGG-**Z**	20-mer	IV	mice
8	ACACCCAATTCTGAAAA⟦TGG⟧	20-mer	IV	mice
9	ACACCCAATTCTGAAA⟦ATGG⟧	20-mer	IV	mice
10	CTCTCGCACCCATCTCTCC ^{T}C GAGAGAGG$_T$C	33-mer	IV	rats
11	CUCUCGCACCCATCTCTCTCCCUUCU	25-mer	IV, oral	rats
12	**CTCT**CGCACCCATCTCTCTC**CTTCT**	25-mer	IV, oral	rats
13	CTCTCGCA**CCCATCTC**TCTCCTTCT	25-mer	IV	rats
14	CTCTCGCACCCAUCUCTCCTTCT	25-mer	IV	rats
15	CTCTCGCA**C**CCAUCUCTCCTTCT	25-mer	IV	rats

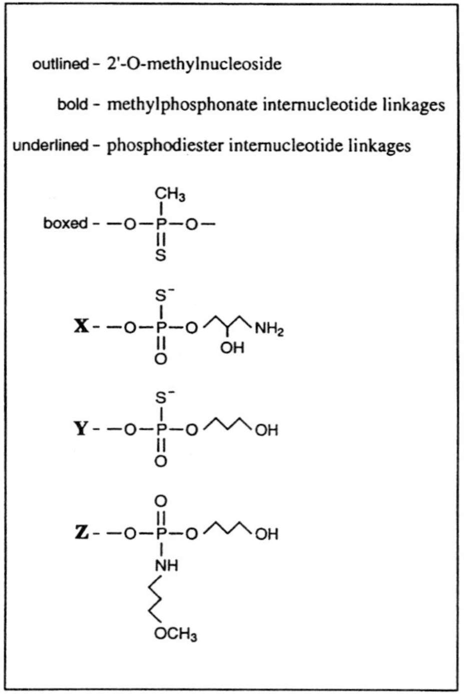

outlined - 2'-O-methylnucleoside

bold - methylphosphonate internucleotide linkages

underlined - phosphodiester internucleotide linkages

boxed -

X -

Y -

Z -

FIG. 3.   Structures of oligonucleotides that have been subjected to pharmacokinetic analysis. IP, intraperitoneal; IV, intravenous; SC, subcutaneous. X, Y and Z represent the chemical group attached to the 3' end, the 5' end, or both the 3' and 5' ends.

The highest concentrations were observed in the liver, kidney, spleen and bone marrow. The major route of elimination was in the urine. Following s.c. administration, the pharmacokinetic profile was similar to that following i.v. administration, except that the peak plasma concentration was lower and the plasma distribution half-life ($T_{1/2}\alpha$) was longer (Agrawal 1995a).

Following i.v. administration of oligonucleotide 2, radioactivity was observed in the contents and tissues of the stomach and the small and large intestines (Zhang et al 1995a). These results led us to investigate its enterohepatic circulation (see below).

The pharmacokinetics of oligonucleotide 2 following i.v. administration in monkeys showed a profile similar to that in mice and rats (Agrawal et al 1995a). In monkeys it was cleared rapidly from the plasma compartment and distributed to various tissues, including kidneys, liver, bone marrow and spleen, where higher concentrations were observed. The major route of elimination was in the urine.

A pharmacokinetic analysis of oligonucleotide 2 carried out in humans following a two-hour i.v. infusion showed that plasma clearance was rapid and the major route of elimination was in the urine (Zhang et al 1995b). Intact phosphorothioate oligonucleotide was detectable in plasma, but not in urine, up to six hours after the end of the infusion.

Degradation

Analysis of extracted phosphorothioate oligonucleotides from plasma and various tissues by HPLC or PAGE showed the presence of both intact phosphorothioate oligonucleotides and their degradation products (Agrawal et al 1991, Sands et al 1994, Zhang et al 1995a). The rate and pattern of degradation of phosphorothioate oligonucleotides were plasma, tissue and time dependent. The ladder of degradation products on PAGE gels suggested that the degradation of phosphorothioate oligonucleotides was primarily by exonucleases. Pharmacokinetics and *in vivo* stability studies of various end-capped phosphorothioate oligonucleotides (oligonucleotides 3–9, Fig. 3) suggested that phosphorothioate oligonucleotides were degraded primarily from the 3' end (Temsamani et al 1993). A recent study in which sequences of the degradation products of phosphorothioate oligonucleotides were identified by sequencing showed that, while the oligonucleotide was degraded primarily from the 3' end, degradation products were also generated from the 5' end or both the 3' and 5' ends (Temsamani et al 1997). In addition to degradation products, we have previously reported the isolation of higher molecular weight metabolites of administered oligonucleotide (Agrawal 1991). Further studies suggest that these high molecular weight metabolites are glucouronic acid conjugates (Temsamani et al 1997).

Impact of phosphorothioate oligonucleotide sequence on pharmacokinetics

Although most phosphorothioate oligonucleotides have similar pharmacokinetic profiles, we have observed significant changes in plasma clearance, tissue disposition, elimination and *in vivo* stability of phosphorothioate oligonucleotides that can form secondary structures, or hyperstructures. A phosphorothioate oligonucleotide (oligonucleotide 10, Fig. 3) that had a secondary structure at the 3' end in the form of a hairpin loop, called a 'self-stabilized oligonucleotide', showed no significant changes in plasma clearance, tissue disposition or elimination, but showed a significant increase in *in vivo* stability (Tang et al 1993, Agrawal et al 1995b, Zhang et al 1995c). Pharmacokinetics of phosphorothioate oligonucleotides containing four contiguous guanosine residues (G_4), which can form hyperstructures, showed significant changes in plasma clearance, tissue disposition, elimination and *in vivo* stability. The position of a contiguous G_4 sequence in a phosphorothioate oligonucleotide is also a factor affecting its pharmacokinetic profile (Agrawal 1997b).

Modulation of phosphorothioate oligonucleotide pharmacokinetics by chemical modification

Pharmacokinetic studies of phosphorothioate oligonucleotides demonstrate that they are degraded by exonucleases, and the degradation products are short lengths of the administered phosphorothioate oligonucleotides. The therapeutic potential of a phosphorothioate oligonucleotide depends on it being present in its intact form. Repeated administration of phosphorothioate oligonucleotides is required to maintain the presence of intact phosphorothioate oligonucleotide *in vivo*, which also increases the presence of the degraded form, and may be additive in causing side-effects.

To control the *in vivo* degradation of phosphorothioate oligonucleotides, we have designed MBOs, which contain appropriately placed segments of phosphorothioate oligonucleotide and segments of modified DNA or RNA. Segments of modified DNA may contain non-ionic internucleotide linkages — e.g. methylphosphonate (Agrawal et al 1990), methylphosphotriester (Iyer et al 1996), phosphoramidate (Devlin et al 1996) linkages — which provide increased nuclease stability and reduced polyanionic nature. Similarly, segments of modified RNA may contain phosphorothioate RNA or 2'-O-alkyl-modified RNA (Metelev et al 1994). Two designs of MBOs have been studied for pharmacokinetics, one in which a modified DNA or RNA segment is placed at both the 3' and 5' ends of the phosphorothioate oligonucleotide (end-modified MBO) (Agrawal et al 1990, Metelev et al 1994) and the second in which the modified DNA or RNA segment is placed in the middle of the phosphorothioate oligonucleotide (centrally modified MBO) (Metelev et al 1994, Agrawal et al 1997a).

Pharmacokinetics of end-modified, mixed-backbone oligonucleotides

Pharmacokinetic studies have been carried out on two end-modified MBOs: one containing four methylphosphonate linkages at both the 3' and 5' ends of the phosphorothioate oligonucleotide oligonucleotide 12 (Fig. 3) (Zhang et al 1996); and one containing four 2'-O-methylribonucleosides at both the 3' and 5' ends of the phosphorothioate oligonucleotide oligonucleotide 11 (Fig. 3) (Zhang et al 1995d, Agrawal et al 1995c). Following i.v. administration, both end-modified MBOs were cleared from the plasma and distributed into tissues more rapidly than oligonucleotide 2. Tissue disposition of end-modified MBOs was similar to that of oligonucleotide 2, but MBOs were retained for longer in tissues. Analysis of the extracted oligonucleotides from plasma and various tissues showed a significant increase in *in vivo* stability of the end-modified MBOs. The major route of elimination of MBOs was in the urine, and this was slower than that of oligonucleotide 2.

The dose-dependent pharmacokinetics of oligonucleotide 11 were also studied. This end-modified MBO was administered to rats at various i.v. doses (i.e. 3.33, 10, 30 and 90 mg/kg). Plasma concentration–time curves showed dose-dependent profiles (Fig. 4). Plasma disappearance curves for the end-modified MBO could be described by a two-compartment model at the four doses, with pharmacokinetic parameters being dose dependent (Table 1). The majority of the radioactivity found in plasma (up to six hours) was associated with the intact end-modified MBO as analysed by HPLC and PAGE. Urinary excretion represented the major pathway of elimination, and was dose dependent. Following administration of 3.33 and 10 mg/kg doses, about 10% of the administered dose was excreted in urine within 24 hours, whereas following

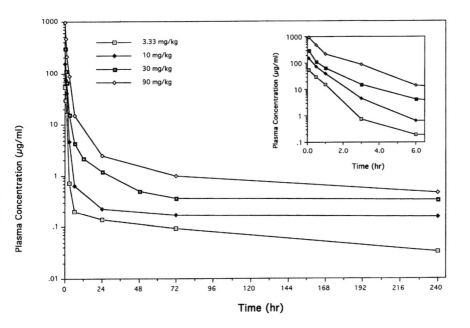

FIG. 4. Plasma clearance of an end-modified, mixed-backbone oligonucleotide (oligonucleotide 11, Fig. 3) following i.v. administration at doses of 3.33, 10, 30 and 90 mg/kg in rats.

TABLE 1 Pharmacokinetic parameters for the end-modified, mixed-backbone oligonucleotide 11 in plasma

Dose (mg/kg)	C_{max} (mg/ml)	$T_{1/2}\alpha$ (hr)	$T_{1/2}\beta$ (hr)	AUC [(µg/ml) hr]	VD_{ss} (ml/kg)	MRT (hr)	CL [ml/(kg hr)]
3.33	59.7	0.46	94.3	63.7	2703.5	52.2	51.8
10	166.1	0.50	97.7	114.4	2940.2	63.0	46.6
30	353.5	0.47	59.7	687.4	2468.3	56.6	43.6
90	996.6	0.61	48.8	1836.2	1847.7	37.7	49.0

The pharmacokinetic parameters of oligonucleotides were estimated using the NLIN procedure of the SAS program (Zhang et al 1995a,c). The selection of models was based on the comparison of the Akaikie's Information Criterion and S.E. of estimated parameters. The two-compartmental model of i.v. bolus injection best fit the plasma concentration–time curve. Values are means based on the experimental data from 30 rats for each dose following administration of ^{35}S-labelled oligonucleotide 11 (Fig. 3). Plasma concentrations were based on the quantitation of radioactivity. AUC, area under the curve; C_{max}, maximum concentration; CL, clearance rate; MRT, mean residue time; $T_{1/2}\alpha$, distribution half-life; $T_{1/2}\beta$, elimination half life; Vd_{ss}, volume of distribution.

administration of 30 and 90 mg/kg, 20–25% of the administered dose was excreted, suggesting saturation of certain body compartments (Fig. 5). Excretion in faeces was higher in rats receiving 3.33 and 10 mg/kg doses than those receiving 30 and 90 mg/kg doses, suggesting changes in the distribution profile at higher doses. The majority of the radioactivity in urine was associated with degradation products of the MBO with lower molecular weights, but the intact MBO was also detected by HPLC and PAGE analyses. A wide tissue distribution of the end-modified MBO was observed in a dose-dependent manner; kidney and liver were the two organs accumulating the highest concentration of the oligonucleotide. The pharmacokinetic parameters of oligonucleotide 11 in the kidney and liver are listed in Table 2. Although concentrations increased in these organs, depending on the administered dose, the percentage of the administered dose taken up by these tissue decreased with increasing doses. Also the proportion of MBO taken up by kidney and liver changed at higher doses (30 and 90 mg/kg) than at lower doses (3.33 and 10 mg/kg).

Pharmacokinetics of centrally modified, mixed-backbone oligonucleotides

We have studied the pharmacokinetics of three centrally modified MBOs: those containing a segment of methylphosphonate, 2'-O-methylribonucleoside phosphorothioate or 2'-O-methylribonucleotide phosphodiester at the centre of a phosphorothioate oligonucleotide (oligonucleotides 13, 14 and 15, respectively, Fig. 3) (Agrawal et al 1997a). We designed the centrally modified MBOs to improve various properties, to enhance the oligonucleotide's potential as a therapeutic agent and to modulate the nature of the metabolites being generated following degradation of the

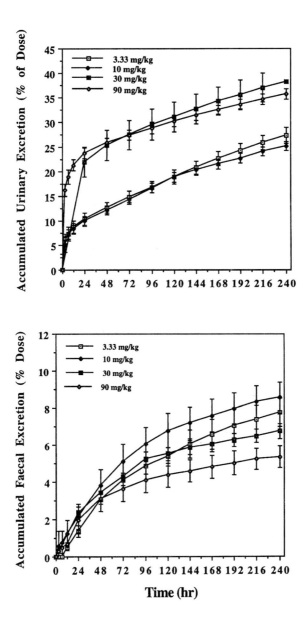

FIG. 5. Accumulated urinary (top panel) and faecal elimination (bottom panel) of an end-modified, mixed-backbone oligonucleotide (oligonucleotide 11, Fig. 3) following i.v administration at doses of 3.33, 10, 30 and 90 mg/kg in rats. Bars represent S.D.

TABLE 2 Pharmacokinetic parameters for the end-modified, mixed-backbone oligonucleotide 11 in tissues

Tissue	Dose (mg/kg)	C_{max} (µg/ml)	T_{max} (hr)	$T_{1/2}A$ (hr)	$T_{1/2}\beta$ (hr)	AUC [(µg/ml) hr]	MRT (hr)	CL [ml/(kg hr)]
Liver	3.33	43.6	8.34	1.41	112.0	7449.6	163.1	0.44
	10	65.7	12.50	2.49	113.3	11699.0	165.9	0.85
	30	67.8	11.03	2.55	126.2	13296.4	183.5	2.26
	90	211.6	19.80	4.06	198.3	65489.1	189.5	1.37
Kidney	3.33	47.8	8.23	1.03	315.5	22194.3	456.4	0.15
	10	119.4	15.67	2.69	250.6	45292.5	363.9	0.22
	30	244.8	38.45	6.04	221.9	86519.7	325.9	0.37
	90	560.1	55.35	20.16	157.2	169967.4	245.6	0.53

The pharmacokinetic parameters of oligonucleotides were estimated using the NLIN procedure of the SAS program (Zhang et al 1995a,c). The selection of models was based on the comparison of the Akaikie's Information Criterion and S.E. of estimated parameters. A first-order absorption, one-compartment model best fit the tissue concentration–time curve. Values are means based on the experimental data from 30 rats for each dose following administration of ^{35}S-labelled oligonucleotide 11 (Fig. 3). Tissue concentrations were based on the quantitation of radioactivity. AUC, area under the curve; C_{max}, maximum concentration; CL, clearance rate; MRT, mean residue time; $T_{1/2}A$, absorption half-life; $T_{1/2}\beta$, elimination half life; Vd_{ss}, volume of distribution.

phosphorothioate oligonucleotide. For example, as discussed above, phosphorothioate oligonucleotides *in vivo* generate degradation products composed of shorter lengths of phosphorothioate oligonucleotide. Centrally modified MBOs degrade to products which have an increasing proportion of modified DNA/RNA segments. Their metabolism results in fragments that either have reduced charges (methylphosphonate linkages), degrade rapidly (2'-O-methylribonucleoside phosphodiester) or are degradation resistant (2'-O-methylribonucleoside phosphorothioate).

Pharmacokinetic analysis of these three centrally modified MBOs showed that plasma elimination was rapid, and the oligonucleotides were distributed to the majority of organs. Analysis of the extracted oligonucleotides showed that degradation products had a distinct profile based on the nature of the central region. In the case of centrally modified MBOs with nuclease-stable methylphosphonate (oligonucleotide 13) and 2'-O-methylribonucleoside phosphorothioate (oligo-nucleotide 14) segments, degradation of the phosphorothioate oligonucleotide was slower at the site of these linkages. Centrally modified MBOs containing 2'-O-methylribonucleoside phosphodiester linkages (oligonucleotide 15) were rapidly digested at the central segment following degradation of the phosphorothioate oligonucleotide. The major route of elimination of these centrally modified MBOs was in the urine and the excretion rate was dependent on the nature of the central region. Centrally modified MBOs containing methylphosphonate (oligonucleotide

13) and 2'-O-methylribonucleoside phosphodiester (oligonucleotide 15) linkages were eliminated more rapidly than centrally modified MBOs containing 2'-O-methyl-ribonucleoside phosphorothioate (oligonucleotide 14) linkages or phosphorothioate oligonucleotides (oligonucleotide 2).

Enterohepatic circulation of oligonucleotides

Pharmacokinetic studies of phosphorothioate oligonucleotides demonstrated that a portion of the administered dose was present in the gastrointestinal tract and its contents following i.v. administration (Zhang et al 1995a,c,d). A limited amount of oligonucleotide-derived radioactivity was excreted in the faeces with a 10-day cumulative excretion being less than 10% for all phosphorothioate oligonucleotides studied. These results indicated that the oligonucleotide-derived radioactivity in the gastrointestinal tract was reabsorbed.

Biliary excretion patterns of a phosphorothioate oligonucleotide (oligonucleotide 2) and an end-modified MBO (oligonucleotide 11) were examined in rats with biliary fistula (Zhao et al 1995). Briefly, following i.v. administration of ^{35}S-labelled oligonucleotides at various doses (10, 20, 30 and 50 mg/kg), bile samples were collected at 5–10 min intervals for four hours, and the radioactivity in bile was quantified. Dose-dependent biliary excretion of oligonucleotide-derived radioactivity was observed (Fig. 6). Peak concentrations of oligonucleotide 2 in bile were reached 20 min after i.v. injection. The peak concentration of oligonucleotide 11, however, was lower than that of oligonucleotide 2 at the same dose level (results not shown). At doses of 10 and 30 mg/kg, peak concentrations were 14.64 ± 1.71 and 47.20 ± 10.13 mg/ml, respectively; 0.92% and 1.04% of the dose was excreted in bile within four hours after i.v. injection at doses of 10 and 30 mg/kg, respectively. HPLC analysis revealed both intact oligonucleotides and their metabolites in the bile.

Oral administration

Further evidence in support of the existence of enterohepatic circulation of oligonucleotides comes from p.o. administration of oligonucleotides in experimental animals. Phosphorothioate oligonucleotides and their end-modified analogues are absorbed through the gastrointestinal tract following p.o. administration. Following oral gavage, the phosphorothioate oligonucleotide 2 was degraded in the contents of the stomach, small and large intestines and showed poor bioavailability of intact phosphorothioate oligonucleotide. The analysis of the extracted oligonucleotide by PAGE from contents of the gastrointestinal tract showed a ladder of degradation products, suggesting that the phosphorothioate oligonucleotide had been degraded by exonucleases. To overcome nuclease digestion, we studied the stability of the end-modified MBO (oligonucleotide 11) following oral administration in rats. This end-modified MBO remained stable in stomach contents, but some degradation was observed in the contents of the small and large intestines (Agrawal et al 1995c,

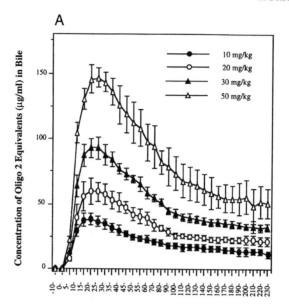

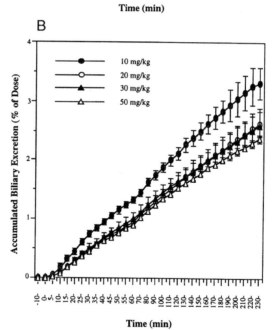

FIG. 6. Concentration (A) and cumulative excretion (B) of a phosphorothioate oligonucleotide (oligonucleotide 2, Fig. 3) in bile following i.v. administration at doses of 10, 20, 30 and 50 mg/kg in rats. Bars represent S.D.

Agrawal & Zhang 1997). Using radioactivity measurements, we calculated that approximately 25% was absorbed from the gastrointestinal tract in the 12 hours following administration of oligonucleotide 11. In further studies, we observed increased absorption following oral administration of oligonucleotide 12, suggesting that a reduction in charge or the number of phosphorothioate linkages may further increase the efficiency of absorption. In recent studies, we administered an end-modified MBO targeted to the RI α subunit of protein kinase A in tumour-bearing mice (xenografts of human breast and colon cancer) and observed an antitumour effect. The antitumour activities of i.p. and p.o. end-modified MBOs were similar (R. Zhang & S. Agrawal, unpublished data 1997).

Plasma clearance and tissue disposition

Traditionally, pharmacokinetic analysis has been based on plasma concentration–time profiles, using various computer-aided programs. Interpretation of pharmacokinetic parameters should consider several important factors including (but not limited to): (1) experimental procedure, e.g. animal species and size, sampling times and size, and observation period; (2) analytical methods, e.g. radioactivity measurements, HPLC, capillary gel electrophoresis and PAGE; (3) the chemical form of the oligonucleotide monitored, e.g. total oligonucleotide vs. free oligonucleotide, and parent oligonucleotide vs. metabolites; and (4) programs for pharmacokinetic analysis, e.g. linear vs. non-linear, different simulation methods, and compartmental vs. non-compartmental. It should be emphasized that the assumption for the use of plasma pharmacokinetic parameters to describe the fate of a given oligonucleotide in the whole body is that the changes in oligonucleotide concentration and chemical form represent the situation in the whole body. This assumption may not always be correct for all classes of therapeutic agents, especially oligonucleotides.

Oligonucleotides are apparently taken up rapidly by various tissues, except the brain, and tissues retain the oligonucleotides for extended periods. No significant qualitative differences in distribution were observed between the phosphorothioate oligonucleotide studied and its modified analogues. Quantitatively, significantly higher retentions of modified oligonucleotides than of phosphorothioate oligonucleotides were seen in all tissues including the liver and kidneys. Further analysis by HPLC and PAGE indicated that the majority of radioactivity in these tissues was associated with intact oligonucleotides, suggesting that: (1) shorter metabolites are cleared more rapidly than intact oligonucleotides; (2) the kidney has an important role in the clearance of oligonucleotides and their metabolites; and (3) the plasma concentrations and degradation profiles of oligonucleotides in plasma do not fully reflect the fate of oligonucleotides in tissues. The selective clearance of intact oligonucleotides vs. their metabolites, as in the case of centrally modified MBOs, may have a significant impact on antisense therapeutic effects, as well as on reducing side-effects caused by metabolites.

Acknowledgements

We thank members of our laboratories who have contributed to work cited in this chapter. Their names appear in the references.

References

Agrawal S 1996 Antisense oligonucleotides: towards clinical trials. Trends Biotechnol 14: 376–387

Agrawal S, Zhang R 1997 Pharmacokinetics of phosphorothioate oligonucleotide and its novel analogs. In: Weiss B (ed) Antisense oligonucleotides and antisense RNA: novel pharmacological and therapeutic agents. CRC Press, Boca Raton, FL, p 57–78

Agrawal S, Mayrand SM, Zamecnik PC, Pederson T 1990 Site-specific excision from RNA by RNase H and mixed phosphate backbone oligodeoxynucleotides. Proc Natl Acad Sci USA 87:1401–1405

Agrawal S, Temsamani J, Tang J-Y 1991 Pharmacokinetics biodistribution and stability of oligodeoxynucleotide phosphorothioates in mice. Proc Natl Acad Sci USA 88:7595–7599

Agrawal S, Temsamani J, Galbraith W, Tang J-Y 1995a Pharmacokinetics of antisense oligonucleotides. Clin Pharmacol 28:7–16

Agrawal S, Temsamani J, Tang J-Y 1995b Self-stabilized oligonucleotides as novel antisense agents. In: Akhtar S (ed) Delivery strategies for antisense oligonucleotide therapeutics. CRC Press, New York, p 105–122

Agrawal S, Zhang X, Lu Z et al 1995c Absorption, tissue distribution and *in vivo* stability in rats of a hybrid antisense oligonucleotide following oral administration. Biochem Pharmacol 50:571–576

Agrawal S, Tan W, Jiang Z, Yu D, Iyer RP 1996 Synthetic methods for the radioisotopic labeling of oligonucleotides. In: Schlingensiepen KH (ed) Antisense oligonucleotide from technology to therapy. Blackwell, Berlin, p 60–77

Agrawal S, Jiang Z, Zhao Q et al 1997a Mixed-backbone oligonucleotides as second-generation antisense oligonucleotides: *in vitro* and *in vivo* studies. Proc Natl Acad Sci USA 94:2620–2625

Agrawal S, Tan W, Cai W, Xie X, Zhang R 1997b *In vivo* pharmacokinetics of phosphorothioate oligonucleotides containing continuous guanosines. Antisense Nucleic Acid Drug Dev 7:245–249

Akhtar S, Agrawal S 1997 *In vivo* studies with antisense oligonucleotides. Trends Pharmacol Sci 18:12–18

Chem TL, Miller P, Ts'O POP, Colvin OM 1990 Disposition and metabolism of oligodeoxynucleoside methylphosphonate following a single IV injection in mice. Drug Metab Dispos 18:815–818

Cossum PA, Sassmor H, Delinger D et al 1993 Disposition of the ^{14}C-labeled phosphorothioate oligonucleotide ISIS 2105 after intravenous administration to rats. J Pharmacol Exp Ther 267:1181–1190

Crooke ST, Graham MJ, Zuckerman JE et al 1996 Pharmacokinetic properties of several novel oligonucleotides analogs in mice. J Pharmacol Exp Ther 277:923–937

Devlin T, Iyer RP, Johnson S, Agrawal S 1996 Mixed backbone oligonucleotides containing internucleotide primary phosphoramidate linkages. Bioorg Med Chem Lett 6:2663–2668

Goodchild J, Zamecnik PC, Kim B 1991 The clearance and degradation of oligodeoxynucleotide following intravenous injection into rabbits. Antisense Res Dev 1:153–160

Iversen PL, Mata J, Tracewell WG, Zon G 1994 Pharmacokinetics of an antisense phosphorothioate oligodeoxynucleotide against *rev* from human immunodeficiency virus type 1 in adult male rat following single injections and continuous infusion. Antisense Res Dev 4:43–52

Iyer RP, Yu D, Jiang Z, Agrawal S 1996 Synthesis, biophysical properties and stability studies of mixed backbone oligonucleotides containing segments of methylphosphotriester internucleotide linkages. Tetrahedron 52:14419–14436

Metelev V, Lisziewicz J, Agrawal S 1994 Study of antisense oligonucleotide phosphorothioates containing segments of oligodeoxynucleotides and 2′-O-methyloligoribonucleotides. Bioorg Med Chem Lett 4:2929–2934

Monia BP, Lesnik EA, Gonzalez C et al 1993 Evaluation of 2′ modified oligonucleotides containing deoxy gaps as antisense inhibitors of gene expression. J Biol Chem 268:14514–14522

Sands H, Gorey-Feret LJ, Cocuzza AJ 1994 Biodistribution and metabolism of internally ^{3}H-labeled oligonucleotides I. Comparison of a phosphodiester and a phosphorothioate. Mol Pharmacol 45:932–943

Tang J-Y, Temsamani J, Agrawal S 1993 Self-stabilized antisense oligonucleotide phosphorothioates: properties and anti-HIV activity. Nucleic Acids Res 21:2729–2735

Temsamani J, Tang J-Y, Padmapriya AA, Kubert M, Agrawal S 1993 Pharmacokinetics biodistribution and stability of capped oligodeoxynucleotide phosphorothioates in mice. Antisense Res Dev 3:277–284

Temsamani J, Roskey A, Chaix C, Agrawal S 1997 In vivo metabolic profile of a phosphorothioate oligonucleotide. Antisense Nucleic Acid Drug Dev 7:159–165

Zamecnik PC 1996 History of antisense oligonucleotides. In: Agrawal S (ed) Antisense therapeutics. Humana Press, Totowa, NJ, p 1–12

Zhang R, Diasio RB, Lu Z et al 1995a Pharmacokinetics and tissue distribution in rats of an oligodeoxynucleotide phosphorothioate (GEM 91) developed as a therapeutic agent for human immunodeficiency virus type-1. Biochem Pharmacol 49:929–939

Zhang R, Yan J, Shahinian H et al 1995b Pharmacokinetics of an anti-human immunodeficiency virus antisense oligonucleotide phosphorothioate (GEM 91) in HIV-1 infected subjects. Clin Pharmacol Ther 58:44–53

Zhang R, Lu Z, Zhang X et al 1995c Stability and disposition of a self-stabilized oligodeoxynucleotide phosphorothioate in rats. Clin Chem 41:836–843

Zhang R, Lu Z, Zhang H et al 1995d In vivo stability and metabolism of a 'hybrid' oligonucleotide phosphorothioate in rats. Biochem Pharmacol 50:545–556

Zhang R, Iyer RP, Yu D et al 1996 Pharmacokinetics and tissues disposition of a chimeric oligodeoxynucleotide phosphorothioate in rats after intravenous administration. J Pharmacol Exp Ther 278:971–979

Zhao H, Lu Z, Diasio RB, Agrawal S, Zhang R 1995 Biliary excretion of oligonucleotides: previously unrecognized pathway in metabolism of antisense oligonucleotides. FASEB J 3:A410

DISCUSSION

Monia: Did you look at the antitumour activity of orally administered end-modified phosphorothioates?

Agrawal: Phosphorothioate oligonucleotides are degraded in the gastrointestinal tract, such that only a small percentage is absorbed as intact oligonucleotide. We have studied end-modified phosphorothioates, and they show antitumour activity following oral administration.

Vlassov: You suggested that the 'extended' oligonucleotides, i.e. those that moved slowly during gel electrophoresis, were conjugated to oligonucleotides in the organisms. Can you expand on this?

Agrawal: In our earlier pharmacokinetic studies, we had observed a product with a slower mobility than the administered oligonucleotide. It seems that this product is readily hydrolyzable by glucourinidase, suggesting it is a glucouronic acid conjugate. However, we do not yet know the location of this residue.

Crooke: We have also looked for conjugates, but we have not found any.

Nicklin: We have observed a peak using mass spectrometry corresponding to the parent compound plus 382 Da. This is believed to represent the slow migrating metabolite.

Crooke: We injected 100 mg/ml, 200 mg/ml or 400 mg/ml ISIS 2302 subcutaneously into rats and monkeys and saw no evidence of toxicity. However, when we injected 200 mg/ml or 400 mg/ml ISIS 2302 subcutaneously into humans (i.e. normal volunteers) we observed substantial lymphadenopathy in the draining lymph nodes, and this effect was clearly concentration dependent because 100 mg/ml did not produce an effect. This shocked us because we had not observed this in our animal studies. Did you observe local lymphadenopathy in immunodeficient patients injected subcutaneously with GEM 91?

Agrawal: We studied healthy volunteers. In some we saw a local irritation and tender/enlarged lymph nodes in the area draining the infection site but the effects were not severe.

Krieg: Stanley Crooke, did you take biopsies of those nodes to see what the cells were?

Crooke: No. It wasn't in the original protocol because we didn't expect to observe these effects.

Rossi: Did you observe nuclear localization of the oligonucleotides in the pharmacology studies?

Agrawal: We've just started doing those studies *in vivo* in mice using fluorescently labelled oligonucleotides, and the preliminary results suggest that oligonucleotides have a different pattern of uptake depending on whether the studies are performed *in vivo* or *in vitro*.

Nicklin: If you administer tritiated phosphorothioate oligonucleotides and do light microscopic autoradiography on the kidney, for example, the oligonucleotide is concentrated in the proximal tubules. Most of it is in the perinuclear area, and there is some nuclear localization.

Rossi: This raises the possibility, in terms of mechanism of action, that when you study something in cell culture you could be looking at something that's totally irrelevant to what will happen in the clinical trial.

Wagner: These studies are extremely difficult to control for because you're using fixed tissue, and it's difficult to control the diffusion of the oligonucleotides in such fixed tissues.

Nicklin: But artefacts due to fixing tend to be artificial nuclear localizations, so the observation that the majority of the oligonucleotides are outside the nucleus suggests that the fixing techniques are fine.

Crooke: In terms of an RNase H mechanism being involved, what are the data that suggest that it is beneficial to have more oligonucleotide in the nucleus? Because the concentrations and activities of RNase H in the nucleus and the cytoplasm are roughly equal.

Matteucci: The question is whether oligonucleotides are trapped in endosomes versus being in the nucleus and cytoplasm, where they can interact with the mRNA.

Crooke: But that accepts the premise that endosomal uptake accounts for the bulk of the distribution of these molecules in animals, and there are no meaningful data that support this. The best approach is to attack the question using as many different methods as possible — such as radiolabelled drugs and autoradiography, rhodomine-labelled drugs or monoclonal antibodies that recognize intact drugs — because every method has some caveats and some limitations. We've used this approach in order to find out which organs, which cells and which subcellular organelles accumulate these drugs in intact animals, and we're finding that the results are to a large extent tissue and cell specific.

Wagner: Could you comment on the clinical studies of GEM 91.

Agrawal: These studies are ongoing in HIV1-infected patients. There are two protocols: in the first GEM 91 is administered by 2 h intravenous infusion; and in the second GEM 91 is administered by continuous intravenous infusion. The former protocol results in doses of up to 3 mg/kg per day, and the latter 4.4 mg/kg per day. The pharmacokinetic and safety profiles at these doses have been established. We have some interesting observations of patients receiving GEM 91 by intravenous infusion for seven days, so we are planning to treat these patients for 14 days.

Matteucci: You talked about the pharmacokinetic parameters of centrally modified oligonucleotides but what was the rationale behind that kind of construct versus the more traditional ones which have modifications at the 5′ and 3′ ends?

Agrawal: The aim of the centrally modified oligonucleotide studies was to improve the efficacy of a given sequence to act as a drug. In centrally modified oligonucleotides the segment of modified DNA or RNA is placed in the centre of the phosphorothioate. The segments of phosphorothioate at the 3′ and 5′ ends serve as substrates for RNase H. Centrally modified oligonucleotides have reduced segments of phosphorothioates. They also show reduced protein binding and complement activation, and prolongation of partial thromboplastin time, compared to phosphorothioate oligonucleotides. Following degradation *in vivo*, centrally modified oligonucleotides generate metabolites that have reduced phosphorothioate segments. We also observe a better safety profile with these oligonucleotides than with phosphorothioates. We have just published these studies (Agrawal et al 1997).

Gait: Surely RNase H cannot cleave at two positions simultaneously? Presumably it is the case that either one or the other site is cleaved.

Agrawal: I am not suggesting that this is the case. When we study RNA cleavage in the presence of centrally modified oligonucleotides and RNase H, we see two cleavage products. The lengths of the cleavage products suggest that they are generated following cleavage at the 5′ and 3′ phosphorothioate segments. The overall rate of RNA cleavage by RNase H is faster than for phosphorothioates.

Gait: Are you saying that the same RNase H molecule binds, cleaves and then moves along and binds and cleaves again, or are there two molecules side-by-side?

Agrawal: It is difficult to determine whether two molecules of RNase H are involved or whether it is the same RNase H molecule that just moves along.

Crooke: Escherichia coli RNase H has a clear binding footprint of about 10 nt. The cleavage unit covers about 4 nt and the binding unit about 6 nt, so it is possible to accommodate one RNase H molecule per helical turn at any given time. Unfortunately, the human RNase H has not yet been cloned, and so the data have to be interpreted with more caution because the enzyme is impure. However, the footprint of that enzyme is larger, probably because it carries some other enzymatic activity. Therefore, whether the footprint of the enzyme can accommodate such an event may just be a statistical matter, i.e. that there are enough RNA molecules and enough protein. Another intriguing situation we've been studying is the effect of oligonucleotide-induced structure on RNase H. Based on what we've seen, it would be surprising if two RNase H molecules could bind simultaneously to this structure and result in cleavage.

Reference

Agrawal S, Jiang Z, Zhao Q et al 1997 Mixed-backbone oligonucleotides as second-generation antisense oligonucleotides: *in vitro* and *in vivo* studies. Proc Natl Acad Sci USA 94:2620–2625

Controversies in the cellular pharmacology of oligodeoxynucleotides

C. A. Stein

Department of Medicine, Columbia University, College of Physicians and Surgeons, 630 W 168 Street, Black Building 20–07, New York, NY 10032, USA

Abstract. Phosphodiester and phosphorothioate oligodeoxynucleotides are polyanions that cannot passively diffuse across cell membranes. Instead, the processes of adsorbtive endocytosis and pinocytosis probably account for the great majority of oligo-deoxynucleotide internalization in most cell types. Oligodeoxynucleotides can adsorb to heparin-binding, cell surface proteins. An example of such a protein is the integrin Mac-1 (αMβ2; CR3; CD11b/CD18), a receptor for fibrinogen which is found on neutrophils, macrophages and natural killer cells. Up-regulation of neutrophil cell surface Mac-1 expression by interleukin 8, arachidonic acid or tumour necrosis factor α leads to increased cell surface oligodeoxynucleotide binding and internalization. Binding and internalization can be blocked by both fibrinogen and by anti-Mac-1 monoclonal antibodies. Subsequent to internalization, oligodeoxynucleotides reside in subcellular vesicular structures, i.e. endosomes and lysosomes. However, in the absence of permeabilizing agents, these compartments may be sites of sequestration and the oligomers may be unavailable for antisense activity. At present, controversy surrounds the use of guanosine-rich phosphorothioate oligodeoxynucleotides as antisense agents. We examined the ability of the 24 mer antisense rel A (p65) phosphorothioate oligodeoxynucleotide to inhibit nuclear translocation of NFκB in K-BALB murine fibroblasts. 7-Deaza-2'-deoxyguanosine substitution in the 5' guanosine quartet region demonstrated that inhibition of nuclear translocation could not be due to a Watson–Crick antisense effect. Rather, we favour the explanation that the parent molecule may be a sequence-specific, apatameric decoy.

1997 Oligonucleotides as therapeutic agents. Wiley, Chichester (Ciba Foundation Symposium 209) p 79–93

The power of the antisense oligodeoxynucleotide technology lies in the ability of these compounds to bind specifically to a target mRNA via Watson–Crick base pair interactions. Clearly, though, in order for this methodology to be successful, the oligodeoxynucleotide must first be internalized by the target cell, and then trafficked to a relevant cellular compartment (probably the nucleus) where hybridization and then (presumably) RNase H cleavage of the mRNA strand can occur. However, the oligodeoxynucleotides in common use today both in tissue culture and in therapeutic clinical trials, i.e. phosphodiesters and phosphorothioates, are polyanions. As such, they cannot diffuse through cell membranes, which indeed represent a formidable barrier to the ultimate success of the technology. This chapter will discuss methods by which cells internalize and compartmentalize oligodeoxynucleotides, and it will describe studies that have elucidated the nature of one of the cell surface oligodeoxy-nucleotide-binding proteins on the surface of human neutrophils. It will also address the question of the 'non-specific' biological effects of guanosine-rich oligodeoxy-nucleotides by examination of the cellular pharmacology of the frequently used antisense rel A (p65) construct.

Adsorbtion of oligodeoxynucleotides to the cell membrane

In order for oligonucleotides to be effective antisense agents, they must first enter cells and achieve the appropriate concentration in the correct intracellular compartment. Bergan et al (1993) overcame the cell membrane barrier to oligodeoxynucleotide transport by electroporation, and showed an increase in the ability of an antisense oligomer to diminish c-*myc* protein synthesis in U-937 human histiocytic lymphoma cells. In the absence of electroporation, this construct was minimally effective. In other experiments, Spiller & Tidd (1995) and Giles et al (1995) permeabilized KYO1 human chronic myelogenous leukaemia cells with streptolysin O and achieved sequence-specific cleavage of *bcr–abl* mRNA. No effect was seen in the absence of permeabilization.

Charged oligodeoxynucleotides are taken up actively by cells, in many cases via a calcium-dependent manner (Stein et al 1993, Wu-Pong et al 1994). The rate of internalization of single-stranded, hairpin and dumbbell decoy phosphodiester oligodeoxynucleotides appears to be similar, at least in T cells (Aguilar et al 1996). The process of internalization is slowed by metabolic inhibitors such as deoxyglucose, cytochalasin B and sodium azide (Yakubov et al 1989). Not surprisingly, the rate of oligomer internalization is also temperature dependent. There appears to be little question that oligodeoxynucleotide internalization in cells, at least in tissue culture, depends predominantly on the two processes of adsorbtive endocytosis and fluid-phase endocytosis (pinocytosis). Adsorbtive endocytosis has been implicated because oligodeoxynucleotides that adsorb well to the cell surface (e.g. phosphodiesters and phosphorothioates) (Stein et al 1993) tend to be internalized to a much higher degree than those oligodeoxynucleotides that do not (methylphosphonates and peptide nucleic acids). Furthermore, if other polyanions (e.g. pentosan polysulfate) are used as competitors of the binding of a charged oligomer to the cell surface, net inter-nalization decreases. As will be discussed later, there are bystander proteins on the cell surface that can adsorb oligodeoxynucleotides. Subsequent to this adsorbtion, the complex is internalized as endosomes form. These bystander proteins are pre-dominately, if not exclusively, heparin-binding proteins, but at the current time there is no evidence for a specific oligodeoxynucleotide receptor of the Brown and Goldstein low density lipoprotein receptor type. Furthermore, we have obtained data that demonstrate that the scavenger receptor does not bind oligodeoxyribonucleotides, as opposed to its ability to bind to at least some oligoribonucleotides (Benimetskaya et al 1997).

The contribution of adsorbtive endocytosis to the internalization process is also suggested by studies of 5' cholesteryl-modified oligodeoxynucleotides (Krieg et al 1993). This modification increases the adsorbtion of the oligodeoxynucleotide via a hydrophobic interaction of the cholesteryl moiety with the cell membrane; net oligomer internalization is also dramatically increased relative to the non-modified oligomer. In addition, mitogen-treated B and T lymphoid cells, which have higher rates of membrane turnover than quiescent cells, also have higher rates of oligomer internalization (Krieg et al 1991). However, in this case, the increase in rate of oligomer internalization is most likely multifactorial, and results not only from an increase in the rate of membrane turnover, but from an increase in cell-surface, heparin-binding (and thus oligodeoxynucleotide-binding) protein content as well.

An additional mechanism of internalization that operates at relatively high oligodeoxynucleotide concentration is pinocytosis, or fluid-phase endocytosis (Vlassov et al 1994). This is the process by which cells constitutively engulf water and dissolved solute from the bulk, or fluid, phase. Although this process is inefficient, it probably accounts for a high percentage of net oligonucleotide internalization, especially when the oligonucleotide concentration is greater than the K_d of oligomer adsorbtion to cell surface heparin-binding proteins. However, regardless of the relative contribution of these processes, the oligomers, despite the precise mechanism of internalization, will enter an endosomal compartment.

At present, there are a large number of cell surface proteins to which oligonucleotides have been claimed to bind. Initially, a protein (or set of proteins) with a molecular mass of 75–80 kDa was discovered (Yakubov et al 1989, Loke et al 1989). This protein has also been called the nucleic acid binding receptor 1 (NABR-1). It was isolated by oligo-dT cellulose chromatography of radioiodinated HL-60 human peripheral blood cell membranes; it was also affinity labelled with an oligonucleotide bound to the Denny-Jaffe reagent, which contains an ^{125}I-labelled aromatic azido group conjugated to an acylating substituent via a cleavable group (Geselowitz & Neckers 1992). NABR-1 is found not only in HL-60 cells, but also in African green monkey COS-1 and Vero fibroblasts, L-671 mouse myoblasts, mouse hepatocytes, and mouse, mink and Chinese hamster ovary (CHO) cells (Vlassov et al 1994). Yet another protein of molecular mass 46 kDa was recognized as a cell surface phosphodiester oligodeoxynucleotide-binding protein on T15 mouse fibroblast cells. (Hawley & Gibson 1996). An oligonucleotide-binding protein (molecular mass 28–34 kDa) appears to be found on lymphoid cells (Goodarzi et al 1991). Others have found that binding of DNA will occur to several proteins (molecular masses 79, 59 and 28 kDa) in normal lymphoid cells (Gasparro et al 1990). The 59 kDa protein may be CD4, which we have shown binds with high affinity to both phosphodiester and phosphorothioate oligodeoxynucleotides (Yakubov et al 1993, Stein et al 1991). More recently, Beltinger et al

(1995) have examined binding of biotinylated phosphorothioate oligomers to K-562 human chronic myelogenous leukaemia cell membranes, and found binding proteins of molecular masses 137–147, 79–85, 43–46, 29–32 and 20–22 kDa. Yao et al (1996) detected two proteins of molecular masses 100 and 110 kDa on the surface of Hep G2 human hepatocellular carcinoma cells. They confirmed that the binding of a 21 mer phosphodiester oligomer could be competed by other polyanions, including phosphorothioates and dextran sulfate. Lower molecular weight digestion fragments were observed in the absence of phenylmethylsulfonyl fluoride (PMSF), a serine protease inhibitor. Rappaport et al (1995) have identified two 20 mer phosphorothioate oligo-deoxynucleotide-binding proteins of molecular masses 46 kDa and 97 kDa, isolated from enriched preparations of purified renal tubular brush border membranes. It is possible that one or both of these proteins acts as an ion channel, and the oligonucleotide carries the current. The 45 kDa protein isolated from renal brush border membrane was purified and reconstituted in an artificial bilayer system that models a cellular membrane. When purified protein is introduced into the bilayer system and oligodeoxynucleotides are added, gated channel activity is observed. Channel activity is not observed in the absence of the 45 kDa protein, in the absence of oligonucleotide or when an unrelated protein is used in the lipid bilayer. The rate at which this putative transport protein can translocate oligonucleotide is unknown at present.

We (Benimetskaya et al 1997) have determined that a major oligodeoxynucleotide-binding protein is the protein Mac-1 (CD11b/CD18; CR3; $\alpha M\beta 2$). This molecule, found predominately on the surface of human neutrophils, natural killer cells and macrophages, is a heparin-binding, pro-adhesive integrin. A radioactive, alkylating phosphodiester 18 mer homopolymer of thymidine bound to purified Mac-1 in a concentration-, Ca^{2+}- and Mg^{2+}-dependent manner. Binding occurred on both subunits of Mac-1 and was best described by a one-site binding model of the Michaelis-Menton type (K_d (αM) = 17 μM; K_d ($\beta 2$) = 8.8 μM). The binding was competitive with other polyanions, including phosphorothioate oligodeoxynucleotides, suramin and a discrete persulfated heparin analogue. Soluble fibrinogen, a natural ligand for Mac-1, was an excellent competitor of the binding of a fluoresceinated phosphorothioate oligodeoxynucleotide (FSdC15) to both tumour necrosis factor α (TNF-α), and activated and non-activated polymorphonuclear leukocytes (PMNs). Treatment of PMNs with substances known to up-regulate cell surface Mac-1 expression (e.g. TNF-α, leukotriene B4, arachidonic acid, interleukin 8 and C5a) increased anti-Mac-1 monoclonal antibody (mAb) cell surface binding, as well as cell surface binding of FSdC15. Oligodeoxynucleotide binding was inhibited by anti-Mac-1 mAbs directed against either the αM or $\beta 2$ chains, and the increase in cell surface binding was correlated with a three- to fourfold increase in FSdC15 internalization by PMNs. Furthermore, CHO cells transfected with both αM and $\beta 2$ chains demonstrate dramatically increased, mAb-sensitive FSdC15 internalization compared to control, non-transfected cells. The binding of phosphorothioate oligodeoxynucleotides to Mac-1 has functional significance, as $\beta 2$-dependent migration through Matrigel was dramatically inhibited by SdC28. In addition, the production of reactive oxygen species (ROS) in PMNs adherent to fibrinogen was dramatically increased by SdC28 in TNF-α- and fMLP (formyl-Met-Leu-Phe)-stimulated PMNs. This increment in ROS production was blocked by an anti-Mac-1 mAb. Thus, these data demonstrate the ability of an integrin to serve as a cell surface oligodeoxynucleotide-binding and internalization protein, and imply that there may be other related molecules on other cell types that can serve an identical function.

Oligodeoxynucleotide internalization and compartmentalization

The rate of internalization of oligonucleotides is affected, at least in HL-60 cells, by inhibitors of protein kinase C activity (Stein et al 1993). Tested inhibitors include staurosporine, the isoquinoline sulfonamide H7 and tamoxifen. Unfortunately, none of these compounds is highly specific for protein kinase C (PKC). The results obtained with more specific PKC inhibitors will be described below. It should be noted that phosphorothioate oligodeoxynucleotides themselves are competitive inhibitors, with respect to substrate binding (in this case a peptide fragment of the epidermal growth factor receptor), of partially purified PKC-β1 isoform, when assayed in the presence of phosphatidylserine cofactor and Ca^{2+}. However, the K_i is 5.4 μM for SdC28; thus it is highly unlikely that this concentration of phosphorothioate oligomer can be attained in the cytoplasm (Khaled et al 1995). Therefore, direct inhibition of PKC activity by phosphorothioate oligomers also does not appear to be attainable, if the *in vitro* data are reasonable models of the *in vivo* situation.

The bulk of the internalized oligomer will be trapped in an intracellular vesicle. The oligodeoxynucleotide concentration in the vesicle will increase dramatically because of the shift of water from it into the cytoplasm, and eventually to the media. Thus, the binding of oligodeoxynucleotides to elements contained in the vesicular membrane will increase. It is possible that rare, stochastic, endosomal rupture can cause transfer of oligodeoxynucleotides into the cytoplasm and hence to the nucleus, but this process has never been demonstrated.

Tonkinson & Stein (1994) studied the compartmentalization of phosphodiester and phosphorothioate oligodeoxynucleotides in HL-60 cells. Oligomers were fluorescently labelled; the authors used the pH quenching of fluorescein fluorescence to determine not only the compartment in which the oligonucleotide resided, but also its intracellular pH. In brief, after loading the cells for 6 h (the time chosen to ensure approximate steady-state conditions) and stripping of cell surface-bound fluorescence with SdC28, the total internalized fluorescence was determined by flow cytometry. Then, the calcium ionophore monensin was added to break down the pH gradient between the endosome/lysosome and the cytoplasm. For phosphorothioate and phosphorodithioate (Tonkinson et al 1994), but not phosphodiester oligomers, a dramatic increase in fluorescein fluorescence was observed after monensin treatment. This increase was due to the abrogation of the low pH quenching of fluorescein fluorescence in the presence of monensin. Furthermore, the site of the pH quenching was the endosome or lysosome. This was demonstrated by treatment of the cells with bafilomycin (Nelson 1991), an antibiotic that specifically inhibits the proton-pumping ability of the H^+-ATPase, the enzyme that acidifies the lumen of the endosome (Yoshimori et al 1991). In the presence of bafilomycin, the signal from FSdT15 was equivalent to that produced by the monensin treatment described above, indicating that no acidification took place. The fact that additional treatment of the bafilomycin-treated cells with monensin did not produce a signal more intense than bafilomycin treatment alone indicated that all of the acidification was occurring in a bafilomycin-sensitive location, i.e. the endosome or lysosome.

Oligonucleotides were found to undergo efflux from the HL-60 cells. The rate of efflux, for all classes of charged oligodeoxynucleotide studied, was best described by:

$$C_T = Ae^{-\alpha t} + Be^{-\beta t} \qquad (1)$$

where C_T is the amount of oligomer remaining internalized in the cell at any time T, α and β are the rate constants of efflux, and A (the amount of material in the α compartment)+B (the amount of material in the β compartment) = 100%. Each exponential component of the sum in Equation 1 is also referred to as a 'compartment'. While this mathematical description does not assign an actual cellular structure to each 'compartment', we know from the studies described above that efflux is actually occurring predominately from the vesicular structures (i.e. endosomes/lysosomes).

We also evaluated the efflux behaviour of 12 phosphodiester oligomers in HL-60 cells (Tonkinson & Stein 1994). These compounds predominately enter a 'shallow' (relatively rapid efflux; short $t_{1/2}$) compartment. The value of A = 61%±4%; the value of α is $10\,h^{-1}$. In contrast, about 36% of the oligonucleotide enters the B, or deep, compartment (relatively slow efflux, long $t_{1/2}$ [$\beta = 0.329\,h^{-1}$]). In contrast, for three phosphorothioate oligonucleotides studies (15–28 mer), the situation is reversed. Only 18% enters the shallow (A) compartment, while 80% enters the B compartment ($\alpha = 3.5\,h^{-1}$; $\beta = 0.131\,h^{-1}$). Acidification of the phosphorothioate and phosphophorodithioate compounds occur in the B compartment, where the average pH values are approximately 6.0 and 5.5, respectively. In contrast, phosphodiester oligonucleotides were not acidified (pH = 7.2). Similar efflux data were obtained with rhodamine-labelled oligonucleotides, indicating that it was probably not the fluorescent group that was responsible for the efflux properties.

Sequestration of the bulk of the oligonucleotides in endosomal/lysosomal compartments occurs not only in HL-60 cells, but in K-562 cells and DU 145 human prostate cancer cells as well. This compartment is, of course, useless for antisense activity, and may thus represent a dead-end for the oligonucleotides. Accordingly, knowing the rate of loss of oligonucleotide from the endosomal/lysosomal compartment into the cytoplasm would be extremely useful. This is particularly so in light of microinjection data (Fenster et al 1994), which suggest that when the oligonucleotide enters the cytoplasm it is rapidly translocated to the nucleus.

One method for determining the rate of loss of oligonucleotide from the vesicular compartment to the cytoplasm is a ratiometric one, which takes advantage of the ability of the dye rhodamine green dextran to co-localize with fluorescently labelled oligodeoxynucleotides in the vesicular compartment. The dye is endosome impermeable. Thus, a change in the ratio of fluorescein fluorescence/dye fluorescence in the endosomes, as determined by confocal microscopy, implies leakage of the oligodeoxynucleotide from endosome into cytoplasm. We have performed this type of ratiometric measurement in DU-145 cells, but we have not observed a change in ratio even after 12 h of incubation, implying that the 'leak rate' is extremely slow. The development of reagents that can increase the 'leak rate' in a controlled manner would seem to be an important undertaking. In fact, recent data suggest that increasing the leak rate (Zabner et al 1995) is the method by which cationic lipids such as lipofectin increase the concentration of oligodeoxynucleotide in the cytoplasm and nucleus.

More recent data indicate that the rate of internalization of oligodeoxynucleotides in diverse cell types is also controlled by serum lipids, particularly those bound to albumin. Albumin contains approximately five binding sites/molecule for lipids. If these lipids are removed by methanol, oligodeoxynucleotide internalization declines precipitously. Moreover, baseline levels of internalization can be reached if the lipid is replaced. The most stimulatory lipids are the omega-3 fatty acids, especially linoleic (150 mM) and arachidonic acids. Oleic acid is not as active as either of these two, and saturated fatty acids, such as stearic and palmitic acids, are entirely ineffective, as are the initial oxidation products of linoleic acid, 9S- and 13S-hydroxyoctadecadienoic acid (HODE). Linoleic acid is a known activator of several isoforms of PKC, including α, $\beta 1$, $\beta 2$, γ and the atypical, Ca^{2+} independent form ζ (Liu 1996). The highly specific PKC inhibitor Ro-318425 (Bit et al 1993) entirely blocked linoleic- and fatty acid-induced increases in oligodeoxynucleotide internalization in K-562, HL-60 and RD human rhabdomyosarcoma cells, as well as in human PMNs. In addition, it also reduced the baseline, unstimulated internalization of fluoresceinated oligodeoxynucleotides. Use of another PKC inhibitor, Go 6976, which is specific only for the Ca^{2+}-dependent PKC isoforms (Martiny-Baron et al 1993), did not affect oligodeoxynucleotide internalization. Examination of the particulate fraction of K-562 cells treated with linoleic acid demonstrated translocation of PKC-ζ to the cell membrane by Western blotting, but of none of the other isoforms present in this cell line.

We then examined the role of linoleic acid and the PKC inhibitor Ro-318425 on the internalization of fluoresceinated oligodeoxynucleotides in K-562 cells. The efflux rate, the efflux parameters A, α and B, β (see above), and the extent of acidification of internalized oligodeoxynucleotide are identical in linoleic acid-treated and untreated cells if the load time is 6 h. Thus, a change in the entrance rate or in the nature of the oligodeoxynucleotide compartmentalization must be present in order to account for a net increase in internalized material. In fact, our preliminary data, obtained at short-loading times, indicate that the role of Ro-318425 is to prevent the shunting of oligodeoxynucleotides from the shallow to the deep intracellular compartments, thus preventing accumulation of material in the deep compartments (see above). Taken together, the evidence indicates that the role of the omega-3 fatty acids is to induce membrane translocation of PKC-ζ, which then acts locally by an as yet unknown mechanism to promote accumulation of oligodeoxynucleotide in deep, acidified compartments. However, linoleic acid treatment of cells does not at this point seem to increase anti-sense efficacy, most likely due to sequestration in these deep compartments.

The question of the guanosine-rich oligodeoxynucleotide

A guanosine-rich oligonucleotide that has been extensively used as an 'antisense reagent' is rel A (p65), which is targeted at the p65 subunit of NFκB. The murine version of this oligomer, which we have studied in the phosphodiester and phosphoro-thioate forms (and whose solution behaviour is virtually identical), is a 24 mer whose sequence is $G^1A^1G^2G^3G^4G^5A^2A^3A^4$.CAG^6 ATC G^{7}TC CAT G^{8}G^{9}T. Although visualization after electrophoresis in native gels reveals a band with the rate of migration of a typical 24 mer, two other, slower moving bands are also observed, one with relative mobility of about 0.9, and the other with relative mobility of about 0.01. This lowest mobility band is almost certainly a tetraplex. The concentration of these bands is independent of metal ion concentration (Na^+, K^+, Cs^+, Mg^{2+}) but the ionic strength dependence is not yet known. After boiling and 7 M urea PAGE only a single band (relatively mobility 1.00) is observed. The band of mobility 0.01 will slowly ($t_{1/2}$ = many hours) disaggregate at 25 °C in solution to form monomeric, mobi-lity 1.00 material. Furthermore, substitution of G^2 by 7-deazaguanosine, which mark-edly reduces guanosine-self association by preventing the formation of Hoogsteen base-pair formation, produces only a single band (mobility = 1.00) in a native gel. Sub-stitution at G^1 causes elimination of the 0.9 band and the formation of another band of relative mobility 0.95. Substitution at G^5 produces no change in electrophoretic mobility from the parent compound.

When the rel A phosphodiester oligomer is labelled on the 5′ terminus with ^{32}P, treated with dimethyl sulfate and piperdine and the cleavage products examined by SDS-PAGE, all of the guanosine residues (G^L–G^5) are clearly observed. However, when the same oligonucleotide is labelled on the 3′ terminus with ddA, G^1 and G^2 (and some-times G^1–G^5) are heavily protected against dimethyl sulfate/piperidine cleavage.

In combination, the data are consistent with the hypothesis that the parent oligo-nucleotide is in equilibrium with a species that is produced by formation of a 5′ loop structure in which G^1 interacts with G^2. In addition, higher-order structures, such as tetraplexes, are also found. A phosphate group at the 5′ position appears to produce sufficient charge–charge repulsion with the backbone to cause the loop to open. This looped oligonucleotide appears to be able to multimerize, although the rate at which this occurs is currently unknown. Nevertheless, the presence in solution of various higher-order oligonucleotide structures whose differential protein-binding capacities are not known makes it virtually impossible to propose appropriate control sequences. If K-BALB (K-ras transformed) murine fibroblasts are treated with the rel A antisense phosphorothioate oligodeoxynucleotide, nuclear p65 (as demonstrated by binding to the NFκB consensus sequence) is virtually eliminated. However, if any of the guanosines at the 5′ terminus of the molecule are substituted by 7-deaza-2′deoxyguanosine, p65 nuclear expression is no longer down-regulated. However, if G^7, which is not involved in Hoogsteen base pair formation, is substituted by 7-deazaguanosine, the level of nuclear p65 expression is identical to what is seen with the parent, unsubstituted antisense p65 oligodeoxynucleotide. Furthermore, we have also examined the effects of these oligodeoxynucleotides on the nuclear expression of the Sp1 transcription factor. Treatment of the K-BALB cells with the parental p65 antisense oligodeoxynucleotide leads to almost total down-regulation of nuclear Sp1 activity. However, this down-regulation was no longer observed when 7-deazaguanosine substitution was performed in the 5' region. Substitution at G^7, however, restored the pattern observed with the parental antisense rel A oligomer, i.e. almost complete down-regulation. Thus, the positional effects of single 7-deazaguanosine substitution demonstrate that the modulation of the nuclear levels of p65 by the rel A antisense phosphorothioate oligodeoxynucleotide cannot be due to Watson–Crick antisense hybridization. Rather, these effects are probably due to an aptameric, sequence-specific 'decoy' effect of one or more of the folded forms on an as yet unknown transcription factor. These results help to clarify some of the confusion surrounding the 'non-specific' effects of guanosine-rich oligodeoxynucleotides, and point the way toward further exploitation of the pleotropic properties of these fascinating molecules.

Acknowledgements

C. A. S. is partially funded by NIH-NCI 60639. He wishes to acknowledge the help by all of his contributors to and collaborators of this work.

References

Aguilar L, Hemar A, Dautry-Varsat A, Blumenfeld M 1996 Hairpin, dumbbell, and single-stranded phosphodiester oligonucleotides exhibit identical uptake in T lymphocyte cell lines. Antisense Nucleic Acid Drug Dev 6:157–163

Beltinger C, Saragovi HU, Smith RM et al 1995 Binding, uptake and intracellular trafficking of phosphorothioate-modified oligodeoxynucleotides. J Clin Invest 95:1814–1823

Benimetskaya L, Loike J, Khaled Z, Wright S, Kai T, Stein CA 1997 Mac-1 (CD11b/CD18) is an oligodeoxynucleotide binding protein. Nat Med 3:414–420

Bergan R, Connell Y, Fahmy B, Neckers L 1993 Electroporation enhances c-*myc* antisense oligodeoxynucleotide efficacy. Nucleic Acids Res 21:3567–3573

Bit R, Davis D, Elliot L et al 1993 Inhibitors of protein kinase C. 3. Potent and highly selective bisindolylmaleimides by conformational restriction. J Med Chem 36:21–29

Fenster SD, Wagner R, Froehler BC, Chin DJ 1994 Inhibition of human immunodeficiency virus type-1 env expression by C-5 propyne oligonucleotides specific for rev response element stem loop V. Biochemistry 33:8391–8398

Gasparro FP, Dall'Amico R, O'Malley M, Heald PW, Elelson R 1990 Cell membrane DNA: a new target for psoralen photoadduct formation. Photochem Photobiol 52:315–321

Geselowitz DA, Neckers LM 1992 Analysis of oligonucleotide binding, internalization, and intracellular trafficking utilizing a novel radiolabeled crosslinker. Antisense Res Dev 2:17–26

Giles RV, Spiller DG, Tidd DM 1995 Detection of ribonuclease H-generated mRNA fragments in human leukemia cells following reversible membrane permeabilization in the presence of antisense oligodeoxynucleotides. Antisense Res Dev 5:23–31

Goodarzi G, Watabe M, Watabe K 1991 Binding of oligonucleotides to cell membranes at acidic pH. Biochem Biophys Res Commun 181:1343–1351

Hawley P, Gibson I 1996 Interaction of oligodeoxynucleotides with mammalian cells. Antisense Nucleic Acid Drug Dev 6:197–206

Khaled Z, Rideout D, O'Driscoll KR et al 1995 Effects of suramin-related and other clinically therapeutic polyanions on protein kinase C activity. Clin Cancer Res 1:113–122

Krieg A, Gmelig-Meyling F, Courley M, Kisch W, Chrisey L, Steinberg AD 1991 Uptake of oligodeoxyribonucleotides by lymphoid cells is heterogeneous and inducible. Antisense Res Dev 1:161–171

Krieg AM, Tonkinson J, Matson S et al 1993 Modification of antisense phosphodiester oligodeoxynucucleotides by a 5'-cholesteryl moiety increases cellular association and improves efficacy. Proc Natl Acad Sci USA 90:1048–1052

Liu JP 1996 Protein kinase C and its substrates. Mol Cell Endocrinol 116:1–29

Loke SL, Stein C, Zhang X et al 1989 Characterization of oligodeoxynucleotide transport into living cells. Proc Natl Acad Sci USA 86:3474–3478

Martiny-Baron G, Kazanietz M, Mischak H et al 1993 Selective inhibition of protein kinase C isozymes by the indolocarbazole Go 6976. J Biol Chem 268:9194–9197

Nelson N 1991 Structure and pharmacology of the proton ATPases. Trends Pharmacol Sci 12:71–75

Rappaport J, Hanss B, Kopp JB et al 1995 Transport of phosphorothioate oligonucleotides in kidney: implications for molecular therapy. Kidney Int 47:1462–1469

Spiller DG, Tidd DM 1995 Nuclear delivery of antisense oligodeoxynucleotides through reversible permeabilization of human leukemia cells with streptolysin O. Antisense Res Dev 5:13–22

Stein CA, Neckers LM, Nair BC, Mumbauer S, Hoke G, Pal R 1991 Phosphorothioate oligodeoxynucleotide interferes with binding of HIV-1 gp120 to CD4. J Acquired Immune Defic Syndr Res 4:686–693

Stein CA, Tonkinson JL, Zhang L-M et al 1993 Dynamics of the internalization of phosphodiester oligodeoxynucleotides in HL60 cells. Biochemistry 32:4855–4861

Tonkinson JL, Stein CA 1994 Patterns of intracellular compartmentalization, trafficking and acidification of 5'-fluorescein labeled phosphodiester and phosphorothioate oligodeoxynucleotides in HL60 cells. Nucleic Acids Res 22:4268–4275

Tonkinson, JL, Guvakova M, Khaled Z et al 1994 Cellular pharmacology and protein binding of phosphoromonothioate and phosphorodithioate oligodeoxynucleotides: a comparative study. Antisense Res Dev 4:269–278

Vlassov VV, Balakireva LA, Yakubov LA 1994 Transport of oligonucleotides across natural and model membranes. Biochim Biophys Acta 1197:95–108

Wu-Pong S, Weiss T, Hunt CA 1994 Calcium dependent cellular uptake of a c-*myc* antisense oligonucleotide. Cell Mol Biol 40:843–850

Yakubov LA, Deeva EA, Zarytova VF et al 1989 Mechanism of oligonucleotide uptake by cells: involvement of specific receptors? Proc Natl Acad Sci USA 86:6454–6458

Yakubov LA, Khaled Z, Zhang L-M, Truneh A, Vlassov V, Stein CA 1993 Oligodeoxynucleotides interact with recombinant CD4 at multiple sites. J Biol Chem 268:18818–18823

Yao GQ, Corrias S, Cheng YC 1996 Identification of two oligodeoxyribonucleotide-binding proteins on plasma membranes of human cell lines. Biochem Pharmacol 51:431–436

Yoshimori A, Yamamoto A, Moriyama Y, Futain M, Tashiro Y 1991 Bafilomycin A1, a specific inhibitor of vacuolar-type H^+-ATPases, inhibits acidification and protein degradation in lysosomes of cultured cells. J Biol Chem 266:17707–17712

Zabner J, Fasbender A, Moninger T, Poellinger K, Welsh M 1995 Cellular and molecular barriers to gene transfer by a cationic lipid. J Biol Chem 270:18997–19007

DISCUSSION

Eckstein: Do linoleic acids improve the antisense effect relative to liposomes? Linoleic acid itself is not toxic so presumably this is one advantage over using liposomes.

Stein: Yes, that's correct. However, what we seem to be doing is increasing the number of vesicles containing oligonucleotides within cells. Therefore, we have solved half of the problem but the other half remains, which is how to get the oligonucleotides out of the vesicles.

Letsinger: Has anyone taken a 'so-called' antisense oligonucleotide, put four guanosine residues in it and found that it's no longer affected?

Stein: I'm not aware that anybody has done this experiment.

Crooke: The question is, why do you think that there's any antisense activity in a molecule that has a G quartet, unless you believe that somehow a fraction of those molecules have chosen not to form a G quartet?

Letsinger: Is the conclusion, therefore, that if there are four guanosine residues in a molecule it won't have antisense activity?

Crooke: The statement is not an absolute one because if there are four guanosine residues in a sequence a G quartet will not form every time. However, if a G quartet is formed it is difficult to imagine that the molecule will have antisense activity.

Gewirtz: The confusion in the literature is that biological effects are observed in molecules that have G quartets, but this is a gross generalization. G quartets are said to have adverse growth effects on haemopoietic cells but we've used G quartet-containing sequences as controls and have observed no biological effect.

Stein: I agree that it depends very much on the context. The NFκB p65 antisense oligonucleotide that I mentioned is a particularly bad one because the guanosine residues are at the 5' terminus. If those guanosine residues were buried somewhere in the middle of the molecule one might not have that much of a problem. It all boils down to finding out what controls the rate of tetrad or tetraplex formation. This has to be studied for every oligonucleotide because of the sequence determinations. It's quite confusing — one can leave these molecules in the freezer for different periods of time and when they are taken out they have different structures.

Monia: Clearly, G quartets can interact with proteins directly but how can you explain the published data that there is a reduction in protein levels (Ratajczak et al 1992)?

Stein: There is definitely a reduction in the levels of protein. We have verified this by Western blotting. We have also verified by Western blotting that if we substitute the guanosine residues in the G quartet region with 7-deaza-2′-deoxyguanosine there is no reduction in p65 protein, whereas there is a dramatic reduction in p65 levels if similar substitutions are made outside the G quartet region. The problem is that NFκB is extremely responsive to the proliferative state of a cell. For instance, it is possible to change the relative ratio of heterodimers to homodimers just by letting the cells grow too densely for too long. If you have a guanosine-rich oligonucleotide that forms tetrads and/or tetraplexes and has a high protein-binding affinity it may bind to a protein on the cell surface, for example, which gives a cell a signal which says 'you're under toxic attack', then it wouldn't surprise me that it is possible to induce nuclear translocation of NFκB.

Monia: Have you tried boiling these G quartet molecules before adding them to the cells?

Stein: No, we haven't boiled the parent p65 antisense NFκB compound.

Crooke: For linoleic acid are you above the critical micelle concentration (CMC)?

Stein: Yes.

Crooke: And are you co-incubating that with the oligonucleotide?

Stein: Yes.

Crooke: So how do you exclude the possibility that you have encapsulated the oligonucleotides in linoleic acid liposomes?

Stein: It's unlikely because the liposome would be highly negatively charged and the oligonucleotides are also negatively charged.

Crooke: But it is possible to trap oligonucleotides in neutral and negative liposomes, and you're above the CMC.

Stein: I can't say it's impossible, but if you put the linoleic acid in the media before you put the oligonucleotides in, and then you wash it out you still get an increase in uptake.

Gewirtz: What is the distribution of PKC-ζ?

Stein: It is relatively ubiquitous, unlike PKC-γ which is predominantly found in neural tissues.

Nicklin: Is it possible that you're creating an acidic microclimate on the cell surface, which would drive cellular association? Because there is a large partition of organic acids in the cell membranes, and we know that acidity drives the cellular association.

Stein: It is possible but I'm not aware of any data which support that. We have not seen increased cellular association *per se* in the presence of linoleic acid, only internalization.

Crooke: Are the internalization experiments performed in the absence of albumin?

Stein: No, they're done in the presence of albumin, in complete media.

Crooke: What's the control that argues against the possibility of the linoleic acid interacting with albumin, which of course it will, thus reducing oligonucleotide binding to albumin, i.e. inhibiting the albumin–antisense interaction?

Stein: The published literature and my own experiments indicate that at the concentrations of phosphorothioates that we use, i.e. 2–5 mM, there isn't an appreciable amount of albumin binding to oligonucleotides.

Crooke: I disagree. Your own experiments show that if you add albumin you inhibit uptake.

Stein: But that may not have anything to do with direct binding of the oligonucleotide to albumin.

Iversen: Acids bind to Site 2, and Site 1 is the primary site for fatty acid interactions. Not much is known about what would happen at an allosteric site. For example, it is possible that at high concentrations the fatty acid alters the conformation at Site 2.

Lebleu: Is it possible that changes in lipid composition affect membrane fluidity?

Stein: Yes, that's likely.

Iversen: Do the PKC data support that? Because it seems likely that downstream events would have little influence on membrane fluidity.

Stein: Yes, the data support it. The increase in oligonucleotide internalization doesn't have to be due to PKC stimulation or to membrane fluidity, rather it could be both.

Nicklin: Have you used a fluid-phase marker, such as ^{14}C-labelled mannitol to assess whether PKC inhibitors are having an effect on the rate of membrane vesicle trafficking?

Stein: No.

Agrawal: Have you looked at different sequences in addition to homopolynucleotides, and have you also studied different cell types?

Stein: Yes we've looked at different sequences but there doesn't seem to be any significant differences between them. We also see essentially the same result in HL-60 human peripheral blood cells, polymorphonuclear leukocytes, K-562 human chronic myelogenous leukaemia cells and RD human rhabdomyosarcoma cells.

Agrawal: Have you tried using the 7-deaza-2′-deoxyguanosine oligonucleotides in animals?

Stein: No we haven't, but it is something that would be of interest to me, particularly in the light of Alan Gewirtz's antisense results using a c-*myb* oligonucleotide. Is there another guanosine residue in your c-*myb* oligonucleotide besides the G_4 sequence?

Gewirtz: That oligonucleotide has a G_4 sequence, but we have used controls for G_4 sequences in the same position and they don't give rise to inhibition. When you modify the guanine in the G_4 sequence you may be modifying the ability of the oligonucleotide to bind to its target. You may also be imparting aptameric effect to the oligonucleotide. In order to demonstrate that the biological effect observed is not an antisense effect you have to demonstrate that the oligonucleotide and its mRNA target no longer form a duplex.

Stein: In my opinion this isn't necessary because if a substitution in the antisense NFκB p65 (rel A) sequence is made at another guanosine residue outside of the G_4 region the protein levels are identical to the parent molecule, whereas if a substitution is made within the G_4 region then the protein levels decrease to 10% of the parent molecule. This is not a statement about any other oligonucleotide, it is a statement only about that one.

Iversen: There is also a phenomenon called purine clash: if the amine is replaced it is possible that the structure of the oligonucleotide is not the same in a G_4 sequence as it would be by itself, i.e. that the bases stack differently for guanosine residues in a run of guanosine residues than they do on their own.

Letsinger: Have you shown that a guanosine residue that far downstream is necessary? What happens if you put in something else, such as a one base mismatch?

Stein: We're in the process of doing that. We have also substituted inosine for guanosine in the G_4 sequence and found that this also destroys antisense activity, no matter which guanosine residue is substituted.

Hélène: What is the evidence that a tetraplex structure is formed?

Stein: They are detected on gels. It is also possible to take the full-length oligonucleotide, which is a 24 mer, and attach it to a shorter one, for example a 13 mer. Five bands can then be seen at the top of the gel, which suggest that it does form tetraplexes, but tetraplexes aren't the only structure which that particular oligonucleotide forms. Clearly, a mixture of structures are formed that are in equilibria because bleeding patterns are observed on the gel.

Hélène: Do these patterns disappear when just one guanosine residue is substituted?

Stein: Yes. All that is left is the monomeric structure.

Toulmé: Is the ability to form the tetraplex dependent on the chemistry of the oligonucleotide?

Stein: We have only done a limited amount of chemistry. For example, we've looked at phosphodiesters and phosphorothioates, and there are some differences between them. We looked at phosphorothioates that were stereoregular only in the G_4 region, and this also seemed to make no difference.

Crooke: Tetraplex structures are observed with peptide nucleic acids, but in this case the higher the number of guanosine residues, the greater the problems with solubility.

Gait: Another chemical reason why you should avoid a run of guanosine residues in your oligodeoxynucleotide is that a purine-rich oligodeoxynucleotide on a polypyrimidine RNA target is much less stable than the other way around. Tom Brown and Andrew Lane have an excellent recent paper on this (Gyi et al 1996).

Agrawal: I would like to mention that when we look at G_4-containing oligonucleotides we find increased uptake in cells (Agrawal et al 1996). When the conditions required for tetraplex formation are changed, cell uptake may also be reduced.

Stein: That's an interesting point. It's possible that 7-deaza-2'-deoxyguanosine may influence some uptake parameter. Also, although I'm not sure why this should happen, tetraplex structures may increase cell uptake and allow the tetraplexes to exit the endosomes. Disaggregation to the monomeric form may then occur.

Crooke: Once phosphorothioate tetraplexes are formed they are stable. They dissociate over a period of days, so it's possible that there are many factors that encourage dissociation. The direct experiments would be to find the tetraplex in the cell and also to look directly at RNA levels.

References

Agrawal S, Iadarola PL, Temsamani J, Zhao QY, Shaw DR 1996 Effect of G-rich sequences on the synthesis, purification, hybridization, cell uptake, and hemolytic activity of oligonucleotides. Bioorg Med Chem Lett 6:2219–2224

Gyi JI, Conn GL, Lane AN, Brown T 1996 Comparison of the thermodynamic stabilities and solution conformations of DNA:RNA hybrids containing purine-rich and pyrimidine-rich strands with DNA and RNA duplexes. Biochemistry 35:12538–12548

Ratajczak MZ, Kant JA, Luger SM et al 1992 *In vivo* treatment of human leukemia in a scid mouse model with c-*myb* antisense oligonucleotides. Proc Natl Acad Sci USA 89:11823–11827

Sequence-specific control of gene expression by antigene and clamp oligonucleotides

Claude Hélène, Carine Giovannangeli, Anne-Laure Guieysse-Peugeot and
Danièle Praseuth

*Laboratoire de Biophysique, Muséum National d'Histoire Naturelle, INSERM U 201,
CNRS URA 481, 43 Rue Cuvier, 75231 Paris Cedex 05, France*

Abstract. Control of gene expression at the transcriptional level can be achieved with
triplex-forming oligonucleotides provided that the target sequence is accessible within
the chromatin structure of cell nuclei. Using oligonucleotide–psoralen conjugates as
probes we have shown that the promoter region of the gene encoding the α subunit of
the interleukin 2 receptor and the polypurine tract of integrated HIV provirus can form
sequence-specific, triple-helical complexes in cell cultures. Oligonucleotide–intercalator
conjugates can inhibit transcription initiation by competing with transcription factor
binding. Oligonucleotide analogues containing $N3'{\rightarrow}P5'$ phosporamidate linkages
form stable triple helices that are able to arrest transcription at the elongation step. A
triple helix can also be formed on a single-stranded target by clamp oligonucleotides. A
clamp targeted to the polypurine tract of HIV RNA is able to block reverse transcription
of the viral RNA.

*1997 Oligonucleotides as therapeutic agents. Wiley, Chichester (Ciba Foundation Symposium 209)
p 94–106*

Gene expression can be controlled with oligonucleotides according to several
strategies (for review see Hélène 1994). Antisense oligonucleotides bind to
complementary sequences on mRNAs and inhibit translation of the message into the
coded protein. Ribozymes are also targeted to mRNAs (or viral RNAs) and induce a
catalytic cleavage of the recognized RNA, thereby inhibiting translation of the mRNA
(or expression of viral RNA functions). An oligonucleotide decoy can be used to
sequester a transcription factor and control the expression of genes that are regulated
by this transcription factor. Several genes are expected to respond to the
oligonucleotide decoy due to the involvement of each transcription factor in the
regulation of gene families. Oligonucleotide aptamers can be targeted to proteins
involved at any step of gene control and expression.

   Control of gene transcription can be achieved with antigene oligonucleotides that
bind to double-helical DNA to form a local triple helix (Thuong & Hélène 1993).

Alternatively, an oligonucleotide may inhibit transcription by strand invasion of a double-helical template, as observed with peptide nucleic acids (PNAs) (Nielsen et al 1994). The targeted sequence may be located in the promoter or enhancer region of the gene or within the transcribed portion. A triple helix can also be formed on a single-stranded nucleic acid by clamp (Giovannangeli et al 1991, 1993, 1996a) or circular (Kool 1991, Wang & Kool 1995) oligonucleotides. If the target is an RNA, oligonucleotides are expected to inhibit translation of a mRNA or reverse transcription of a viral RNA. This chapter will deal with triple helix-forming oligonucleotides and their gene regulatory activities at both the transcriptional and translational levels.

Triple helix formation

Triple helix formation involves the recognition of Watson–Crick base pairs by hydrogen bonding interactions within the major groove of the double helix (Thuong & Hélène 1993). Oligonucleotides and oligonucleotide analogues can wind around the double helix in an orientation that is dependent on the base sequence. Recognition of the purines in T•A and C•G base pairs may be achieved by T and protonated C (C^+), respectively, by forming Hoogsteen hydrogen bonds (as originally described by Hoogsteen 1963). Pyrimidic oligonucleotides adopt a parallel orientation with respect to the oligopurine sequence. The two base triplets T•A × T and C•G × C^+ are isomorphous, i.e. the oligopyrimidine winds without any distortion of its backbone around the targeted double-helical sequence. The requirement for cytosine protonation to form a stable C•G × C^+ base triplet makes the stability of triple helices pH specific for (C,T)-containing oligonucleotides. However, triple helices can be observed at pH 7 if most cytosines have thymine and no cytosine neighbours.

Alternatively, the purines of T•A and C•G base pairs can be recognized by A and G, respectively. A purinic oligonucleotide binds in an antiparallel orientation with respect to the target oligopurine sequence. The two base triplets T•A × A and C•G × G are not isomorphous; therefore, an adjustment of the backbone conformations is required to form a triple helix.

The parallel orientation of (T,C)-containing oligonucleotides and the antiparallel orientation of (A,G)-containing oligonucleotides assume that all nucleotides adopt an anti conformation (Beal & Dervan 1991). Such orientations have been experimentally observed in all experiments reported to date. A syn conformation of the nucleosides would lead to a reverse orientation. It should be noted that T and C^+ can form base triplets with T•A and C•G base pairs, respectively, in a reverse Hoogsteen configuration that should lead to an antiparallel orientation of the (T,C)-containing third strand. This has never been observed with natural oligonucleotides, probably because T•A × T and C•G × C^+ base triplets are not isomorphous and the free energy of base triplet formation and stacking may be less favourable in the reverse Hoogsteen configuration.

Oligonucleotides synthesized with guanosine and thymine residues can also form triple helices with an oligopyrimidine•oligopurine sequence of double-helical DNA. The orientation of the (G,T)-containing oligonucleotide depends on base composition (number of 5′ GpT 3′ and 5′ TpG 3′ steps, and lengths of G and T tracts). Parallel and antiparallel orientations involve Hoogsteen and reverse Hoogsteen configuration of the C•G × G and T•A × T base triplets, respectively (Sun et al 1996).

In order for the third strand oligonucleotide to wind smoothly around the major groove of the DNA double helix, all purines of the target sequence must be on the same strand. Otherwise, the backbone of the third strand would have to cross the major groove at the site where a pyrimidine interrupts a polypurine tract. However, it is possible to recognize a pyrimidine within a polypurine sequence by bases forming a single hydrogen bond with the pyrimidine base. This possibility has been exemplified by introducing a guanosine in a (C,T)-containing oligonucleotide to recognize a thymine in an oligopurine sequence, forming a non-canonical A•T × G base triplet. The energetics of this interaction depend on the flanking base triplets (Kiessling et al 1992). It is also possible to enhance the binding energy by attaching an intercalating agent at the site facing the interruption of the oligopurine sequence (Zhou et al 1995).

The recognition of two oligopurine sequences alternating on the two strands of the DNA double helix can be achieved by two oligonucleotides linked to each other by a linker, whose length and nature depend on the bases in the third strand and the site (5′ PuPy 3′ or 5′ PyPu 3′) where the third strand crosses the major groove (Sun 1995).

The antigene strategy

The specificity of double-helical target recognition by an oligonucleotide provides the basis of an interesting strategy to inhibit gene expression at the transcriptional level (Hélène 1991, 1994). When the target is located within the control region (promoter or enhancer) the bound oligonucleotide may inhibit transcription factor binding. When the oligonucleotide binds downstream of the transcription start site it may inhibit the elongation step of the transcription process.

The possibility of inhibiting transcription by a (G,T)-containing oligonucleotide was first described in an *in vitro* transcription system (Cooney et al 1988). Binding of an oligonucleotide to a transcription factor-binding site competes with the binding of the regulatory protein and, thereby, modulates transcription initiation (Cooney et al 1988, Maher et al 1992, Grigoriev et al 1992, Ing et al 1993). Several *in vitro* transcription systems have been used to demonstrate this competitive inhibition. However, the elongation process is much more difficult to inhibit because the stability of the triple-helical complex is usually not sufficient to arrest the transcription machinery once it is launched on its double-helical template. Two strategies have been described to achieve such a transcriptional arrest.

(1)   The oligonucleotide can be covalently attached to an intercalating agent. The oligonucleotide–intercalator conjugate binds more tightly to its target DNA

due to the additional binding energy provided by intercalation at the triplex–duplex junction or within the triple-helical region (Giovannangeli et al 1996b).

(2) Chemical modifications of the oligonucleotide may provide the analogue with a tighter binding affinity. PNAs do bind tightly to double-helical DNA but they involve a strand-displacement reaction where two PNA molecules bind to one strand (an oligopurine sequence) of the double helix forming a local triple helix, whereas the second strand (an oligopyrimidine sequence) remains single stranded (Nielsen et al 1994). Several other chemical modifications have been tested for their ability to form triple helices (Escudé et al 1993). 2'-O methyl pyrimidine oligonucleotides form more stable complexes than DNA and RNA oligonucleotides. A (C,T)-containing RNA binds more tightly than the corresponding DNA oligonucleotide to a DNA double helix. In contrast, neither 2'-O methyl nor RNA purine oligonucleotides form stable triple helices when compared to a DNA oligonucleotide. Among all chemical modifications tested so far, N3'→P5' phosphoramidate linkages confer upon pyrimidine oligonucleotides a tighter binding than that of isosequential phosphodiester oligomers (Escudé et al 1996). Purine oligophosphoramidates do not appear to form stable triple helices.

Sequence specificity of transcription inhibition by antigene oligonucleotides

Oligonucleotide–intercalator conjugates, PNAs and oligophosphoramidates have been shown to inhibit transcription *in vitro* in a sequence-specific manner. The data available within cells are less abundant. Few studies provide evidence that the effect observed on gene expression is due to oligonucleotide binding to DNA to form a triple-helical complex. In some cases the observed effect on transcription might be due to binding of the oligonucleotide to a transcription factor rather than to DNA. When the gene of interest is carried by a plasmid vector it is possible to provide evidence for the involvement of triple helix formation in the inhibition of gene transcription by introducing mutations in the target sequence. For example, in the promoter region of the α subunit of the interleukin 2 receptor (IL-2Rα) there is a 15 bp oligopyrimidine•oligopurine sequence which overlaps the binding sites of two transcription factors (NFκB and SRF). When a pyrimidine oligonucleotide tethered to an intercalating agent binds to this sequence *in vitro* the triple-helical complex inhibits the binding of the transcription factors (Grigoriev et al 1992). After transfection of the plasmid in lymphocytic cell lines, the oligonucleotide–intercalator conjugate inhibits transcription of a reporter gene (encoding chloramphenicol acetyltransferase) placed downstream of the IL-2Rα promoter. A mutant of the target sequence was constructed in which three pyrimidines interrupted the oligopurine target sequence. This mutant sequence was unable to form a triple-helical complex with the oligonucleotide–intercalator conjugate but the location of the mutation was such that it did not inhibit transcription factor binding. When the mutant plasmid was transfected in the same cell line as the wild-type plasmid, the oligonucleotide–intercalator conjugate did not exhibit any inhibitory effect on transcription from the mutated IL-2Rα promoter as compared to the wild-type sequence (Grigoriev et al 1993). This experiment demonstrated that the effect of the oligonucleotide–intercalator conjugate was indeed due to binding to the targeted DNA sequence in the IL-2Rα promoter and not to another cellular component (e.g. a transcription factor) involved in controlling transcription from this promoter.

In most cases, however, and especially for an endogenous gene, it is difficult to construct a mutant of the target sequence. Control experiments rely upon modifications of the oligonucleotide sequence, and are therefore subject to problems associated with this type of control because all potential interactions of the oligonucleotide (including self association) may change upon sequence alteration.

Accessibility of DNA in cells

One of the main questions raised by the development of the antigene strategy *in vivo* deals with the accessibility of the target sequence within the chromatin structure in the cell nucleus. In order to answer this question we have developed a strategy based upon using oligonucleotide–psoralen conjugates. When such a conjugate forms a triple-helical complex with DNA, the psoralen moiety can be covalently linked to one or both strands of the double helix upon u.v. irradiation (Takasugi et al 1991). The cross-link arrests DNA replication when a restriction fragment containing the target sequence is used as a template for exponential (PCR) or linear amplification using primers flanking the target sequence. In linear amplification using a single primer a truncated product is obtained when replication is arrested at the cross-linked site. If the PCR is used the inhibitory effect of the cross-link can be quantitated by using quantitative PCR methods. Alternatively, if the site of triple helix formation and cross-linking overlaps a restriction site it is possible to reveal the inhibition of restriction enzyme cleavage at this particular site by using probes that overlap the targeted DNA region. The absence of inhibition at other restriction sites for the same enzyme provides an internal control of the sequence specificity of the cross-linking reaction and, therefore, of triple helix formation. The first strategy (linear amplification) has been used for a plasmid vector carrying the IL-2Rα promoter sequence (Guieysse et al 1996). The second (PCR) and third (cleavage inhibition) strategies have been used to demonstrate the accessibility of the proviral HIV sequence in chronically infected cells (Giovannangeli et al 1997). About 30% of the target sequences were shown to be cross-linked by an oligonucleotide–psoralen conjugate after cell permeabilization to increase penetration of the oligonucleotide (as compared to 95% when deproteinized chromosomal DNA was used as a target *in vitro*). The yield of the reaction is expected to depend on several factors, such as cell cycle, irradiation conditions, oligonucleotide concentration and penetration, and nucleosome positioning. Altogether, the present results demonstrate that DNA sequences which have been used as targets for oligonucleotide–psoralen conjugates are indeed accessible within the chromatin structure of cell nuclei. However, this may not be true of all targeted sequences, due to the nucleosomal structure of chromatin. If the oligonucleotide interacts with a sequence to which transcription factors bind to activate transcription it is likely that the oligonucleotide may have access to its target sequence in a similar fashion to transcription factors. Kinetic parameters might play an important role inasmuch as some triple-helical complexes exhibit a slow rate of formation as compared to protein binding (Maher et al 1990, Rougée et al 1992). The targeted sequence in HIV proviral DNA is located in a transcribed region. Therefore, accessibility is not limited to control regions (promoters or enhancers) of genes.

Oligonucleotide–psoralen conjugates have been used to induce site-specific mutations in plasmids. These mutations are located at the specific site where psoralen cross-linking is induced by u.v. irradiation after triple helix formation. They clearly indicate that the target site has been reached by the oligonucleotide within cells. However, until now the target sites have been limited to plasmid vectors and not to endogenous genes. The yield of mutations (less than 10%) reflects only a fraction of the cross-linked sites because it is expected that DNA repair systems remove part of the cross-links to restore the original sequence of DNA (Wang et al 1995, Sandor & Bredberg 1994, Raha et al 1996).

Clamp oligonucleotides

An oligopurine sequence on a single-stranded nucleic acid can be recognized by a complementary (antisense) oligonucleotide. The short double helix with an oligopyrimidine–oligopurine sequence can, in turn, be recognized by a third strand oligonucleotide to form a triple helix. The two oligonucleotides can be linked to each other to form a unique molecule that can clamp the target sequence on the single-stranded template (Giovannangeli et al 1991). The nature of the third strand (oligopyrimidine, oligopurine or (G,T)-containing oligonucleotide) determines its orientation with respect to the oligopurine target sequence. Therefore, the linker between the antisense and the third strand portions will join a 3′ end to a 3′ end or a 5′ end to a 5′ end (for an antiparallel orientation of the third strand), or a 3′ end to a 5′ end (for a parallel orientation). In the latter case, a circular oligonucleotide can be synthesized (Kool 1991). For (mostly) entropic reasons the circular oligonucleotide will bind more tightly than the clamp oligonucleotide, which in turn binds much more tightly than two separate oligonucleotides (at least in the micromolar range of concentrations).

Clamp oligonucleotides have been shown to inhibit DNA primer extension on a single-stranded template under conditions in which antisense oligonucleotides are devoid of any inhibitory activity (Giovannangeli et al 1993). They can also arrest reverse transcription on a single-stranded RNA template. We have recently shown that a clamp oligonucleotide is able to block HIV infection of CD4[+] cells at an early step after infection, most likely at the stage of reverse transcription. No proviral DNA was detected after viral infection. Control experiments were carried out with a modified sequence of the clamp oligonucleotide and, more importantly (see above for antigene oligonucleotides), with a mutated version of the target sequence using the same clamp oligonucleotide. In both cases no inhibition of viral infection was observed, indicating that the inhibitory effect on the wild-type sequence is likely due to clamp formation. (An antisense oligonucleotide targeted to the same sequence exhibited no inhibition.)

Clamp oligonucleotides can be covalently attached to an intercalating agent. If the antisense portion is made a little longer than the third strand portion, intercalation can lock the complex in place on the single-stranded target (Giovannangeli et al 1993).

Conclusion

Triple helix formation represents an alternative to antisense oligonucleotides to control gene function. Antigene oligonucleotides targeted to the DNA double helix can inhibit transcription. Clamp oligonucleotides targeted to a viral sequence can inhibit reverse transcription. They might also inhibit translation of a messenger RNA (even though there are currently no data available on translation inhibition). Circular oligonucleotides forming a triple helix with a single-stranded template might also be useful in both approaches. Strand displacement reactions as observed with PNAs (which involve triple helix formation by two PNAs on one of the two strands) might represent alternatives to antigene oligonucleotides that bind to DNA without any opening of the double helix. Further experiments with oligonucleotide analogues that form stable triple helices (e.g. oligophosphoramidates) will tell us whether the antigene or clamp strategies can be applied to biologically relevant *in vivo* situations.

References

Beal PA, Dervan PB 1991 Second structural motif for recognition of DNA by oligonucleotide-directed triple-helix formation. Science 251:1360–1363

Cooney M, Czernuszewicz G, Postel EH, Flint SJ, Hogan ME 1988 Site specific oligonucleotide binding represses transcription of the human c-*myc* gene *in vitro*. Science 241:456–459

Escudé C, François JC, Sun JS et al 1993 Stability of triple helices containing RNA and DNA strands: experimental and molecular modeling studies. Nucleic Acids Res 21:5547–5553

Escudé C, Giovannangeli C, Sun JS et al 1996 Stable triple helices formed by oligonucleotide N3′→P5′ phosphoramidates inhibit transcription elongation. Proc Natl Acad Sci USA 93:4365–4369

Giovannangeli C, Thuong NT, Hélène C 1993 Oligonucleotide clamps arrest DNA synthesis on a single-stranded DNA target. Proc Natl Acad Sci USA 90:10013–10017

Giovannangeli C, Montenay-Garestier T, Rougée M, Chassignol M, Thuong NT, Hélène C 1991 Single-stranded DNA as a target for triple helix formation. J Am Chem Soc 113:7775–7777

Giovannangeli C, Sun JS, Hélène C 1996a Nucleic acids: supramolecular structures and rational design of sequence-specific ligands. In: Atwood JL, Davies JED, Macninol DD, Vogtle F (eds) Comprehensive supramolecular chemistry. Pergamon Press, New York, p 177–191

Giovannangeli C, Perrouault L, Escudé C, Thuong NT, Hélène C 1996b Specific inhibition of *in vitro* transcription elongation by triplex-forming oligonucleotide–intercalator conjugates targeted to HIV proviral DNA. Biochemistry 35:10539–10548

Giovannangeli C, Diviacco S, Labrousse V, Gryaznov SM, Charneau P, Hélène C 1997 Accessibility of nuclear DNA to triplex-forming oligonucleotides: the integrated HIV1 provirus as target. Proc Natl Acad Sci USA 94:79–84

Grigoriev M, Praseuth D, Robin P et al 1992 A triple helix-forming oligonucleotide–intercalator conjugate acts as a transcriptional repressor via inhibition of NFκB binding to IL-2 receptor α-subunit regulatory sequence. J Biol Chem 267:3389–3395

Grigoriev M, Praseuth D, Guieysse AL et al 1993 Inhibition of interleukin-2 receptor α-subunit gene expression by oligonucleotide-directed triple helix formation. C R Acad Sci Ser III Sci Vie 316:492–495

Guieysse AL, Praseuth D, Grigoriev M, Harel-Bellan A, Hélène C 1996 Detection of covalent triplex inside human cells. Nucleic Acids Res 24:4210–216

Hélène C 1991 The anti-gene strategy: control of gene expression by triplex forming-oligonucleotides. Anticancer Drug Des 6:569–584

Hélène C 1994 Control of oncogene expression by antisense nucleic acids. Eur J Cancer 30A:1721–1726

Hoogsteen K 1963 The crystal and molecular structure of a hydrogen-bonded complex between 1 methylythymine and 9 methyladenine. Acta Crystallogr 16:907–916

Ing NH, Beekman JM, Kessler DJ et al 1993 *In vivo* transcription of a progesterone-responsive gene is specifically inhibited by a triplex-forming oligonucleotide. Nucleic Acids Res 21:2789–2796

Kiessling LL, Griffin LC, Dervan PB 1992 Flanking sequence effects within the pyrimidine triple-helix motif characterized by affinity cleaving. Biochemistry 31:2829–2834

Kool ET 1991 Molecular recognition by circular oligonucleotides: increasing the selectivity of DNA binding. J Am Chem Soc 113:6265–6266

Maher III LJ, Dervan PB, Wold B 1990 Kinetic analysis of oligodeoxyribonucleotide-directed triple helix formation on DNA. Biochemistry 29:8820–8826

Maher III LJ, Dervan PB, Wold B 1992 Analysis of promoter-specific repression by triple-helical DNA complexes in a eukaryotic cell-free transcription system. Biochemistry 31:70–81

Nielsen PE, Egholm M, Buchardt O 1994 Peptide nucleic acids (PNA). A DNA mimic with a peptide backbone. Bioconjugate Chem 5:3–7

Raha M, Wang G, Seidman MM, Glazer PM 1996 Mutagenesis by third strand-directed psoralen adducts in repair-deficient human cells: high frequency and altered spectrum in a xeroderma pigmentosum variant. Proc Natl Acad Sci USA 93:2941–2946

Rougée M, Faucon B, Mergny JL et al 1992 Kinetics and thermodynamics of triple helix formation: effects of ionic strength and mismatches. Biochemistry 31:9269–9278

Sandor Z, Bredberg A 1994 Repair of triple helix-directed psoralen adducts in human cells. Nucleic Acids Res 22:2051–2056

Sun JS 1995 Rational design of switched triple helix-forming oligonucleotides: extension of sequences for triple helix formation. In: Pullman B, Jortner A (eds) Modelling of bimolecular structures and mechanisms. Kluwer Academic, Netherlands, p 267–288

Sun JS, Garestier T, Hélène C 1996 Oligonucleotide directed triple helix formation. Curr Opin Struct Biol 6:327–333

Takasugi M, Guendouz A, Chassignol M et al 1991 Sequence-specific photo-induced cross-linking of the two strands of double-helical DNA by a psoralen covalently linked to a triple helix-forming oligonucleotide. Proc Natl Acad Sci USA 88:5602–5606

Thuong NT, Hélène C 1993 Sequence-specific recognition and modification of double-helical DNA. Angew Chem Int Ed Engl 32:666–690

Wang S, Kool T 1995 Relative stabilities of triple helices composed of combinations of DNA, RNA and 2′-O-methyl-RNA backbones: chimeric circular oligonucleotides as probes. Nucleic Acids Res 23:1157–1164

Wang G, Levy DD, Seidman MN, Glazer PM 1995 Targeted mutagenesis in mammalian cells mediated by intracellular triple helix formation. Mol Cell Biol 15:1759–1768

Zhou BW, Pugas E, Sun JS, Garestier T, Hélène C 1995 Stable triple helices formed by acridine-containing oligonucleotides with oligopurine tracts of DNA interrupted by one or two pyrimidines. J Am Chem Soc 117:10425–10428

DISCUSSION

Eckstein: What is the explanation for the increased T_m in the case of the 3′ phosphoramidates?

Hélène: I can list a number of possible reasons. One is that the single-stranded oligophosphoramidate has a conformation similar to that of RNA: the sugar is in the C3′-endo conformation. This was demonstrated by David Wilson using nuclear magnetic resonance (Ding et al 1996). Therefore, it might be more adapted to triple helix formation than phosphodiesters. Hydration is also one of the major parameters. We have not yet completed the thermodynamic investigations of these systems, but it is likely that the entropic contribution is high.

Southern: To what extent do the intercalators interact with double-stranded DNA?

Hélène: They all interact with double-stranded DNA. The difference in melting temperature is a measure of the difference in binding affinity to the triplex versus the duplex. You start with a double helix with the third strand oligonucleotide bound, and there is an equilibrium with the free double helix. The double helix is still there at the end of the melting process. The increase in T_m of the triplex-to-duplex transition (ΔT_m) measures the higher affinity of the intercalator for the triplex compared to the duplex. Nevertheless, all the intercalators bind to duplexes, although the five-membered ring intercalates weakly and does not stabilize the duplex. The next step in the melting process involves the separation of the two strands of the double helix: you can measure the ΔT_m when the double helix melts, and from this you can determine whether the intercalating agent binds to the double helix. Therefore, we measured both the melting of the triplex to the duplex, and then that of the duplex to single-stranded DNA.

Inouye: How does a mismatch affect the result?

Hélène: We have obviously not investigated all possibilities. However, if intercalation occurs within the triple helix then one should be careful with the mismatch, i.e. the mismatch might be as stable as the targeted sequence. There are fewer problems at the extremities, however, because a mismatch in the triplex at the ends will result in the loss of strong intercalation. Therefore, discrimination may be higher.

Vlassov: Do you have any experimental data on the repair of the cross-linked DNA?

Hélène: There are conflicting reports in the literature. For example, Bernard Lebleu's lab reported that cross-links on plasmid vectors in Chinese Hamster Ovary (CHO) cells can be repaired efficiently (Degols et al 1994). However, in the interleukin-2 receptor

(IL-2R) system that we investigated earlier, we still observed cross-links 72 h after irradiation within the cells. A recent publication also describes the absence of repair of a psoralen adduct in cells after triplex-directed cross-linking (Musso et al 1996).

Lebleu: These situations are different because you looked at the transcription initiation pathway, whereas we looked at transcription elongation.

Hélène: We also did some experiments with the polypurine sequence of HIV which is located in a transcribed region, and we did not see any repair. But we know that repair can take place because mutations are found at the duplex–triplex junction where the psoralen has been cross-linked. Therefore, repair systems are triggered by these covalent triple helices, but this process may not be sufficient to remove all the cross-links. Also, the mutations may arise as a result of monoadduct formation rather than cross-links between the two strands.

Vlassov: Do you believe that the repair process explains the difference in efficiency of cross-linking measured at different times after irradiation?

Hélène: No. In that particular case we irradiated for 30 min and then we extracted the DNA.

Wagner: Do you know if the 30% of cells that had cross-links were in S phase?

Hélène: We did not use synchronized cell cultures. A 30% yield might represent 30% of the cells that are in a cell cycle phase where their DNA is accessible. However, we only observe 50% cross-linking on naked DNA under the same conditions. We could probably increase the cross-linking yield if we irradiated for a longer period of time but this would kill the cells.

Gait: How does the position and type of intercalator affect the cellular uptake of oligonucleotides?

Hélène: We have limited data on this. In some cell types we observed a higher degree of oligonucleotide internalization when it is linked to an intercalating agent such as acridine. We know that the intercalator is not driving the system because a proflavin intercalating agent by itself, for example, heads straight for the nucleus, whereas when it is attached to the oligonucleotide a punctate pattern of fluorescence is observed, suggesting that it is contained within endocytic vesicles.

Gait: What is the reason for the increased uptake in the presence of the intercalating agent? Could another receptor on the cell surface be involved?

Hélène: We don't know. It may depend on the oligonucleotide sequence or the nature of the intercalating agent.

Wickstrom: Was there any cleavage of the plasmid carrying the triplex-forming oligonucleotide, i.e. was there any indication of a triplex-recognizing, DNA-cutting enzyme?

Hélène: We are currently doing some work on this but we don't have any clear-cut results at present. In some cases we observed cleavage of DNA but this result was not reproducible. We have investigated the *Escherichia coli* UvrA/B/C repair system, and showed that when a cross-link is formed in the triple helix in order to target the cross-linking to a specific site, the UvrA/B/C system is partly inhibited if the triplex is still formed. In contrast, if the oligonucleotide is degraded then the psoralen is recognized by the UvrA/B/C system. The system is asymmetrical because the repair machinery can act on either strand. The oligonucleotide prevents binding on one side of the cross-link but does not prevent binding on the other side, and therefore the cleavage takes place at about 6 nt on the 5' side and 22 nt on the 3' side on one strand and not on the other one. Clearly, the repair system can still operate on these oligonucleotide–psoralen conjugates, but when the psoralen is not present we don't see any cleavage. There is one report by Peter Glazer's group at Yale, which shows that in the presence of triple helices that are stable enough to inhibit transcription, repair associated with the inhibition of transcription occurs, so that mutations in the plasmid vectors are observed, even without cross-linking the oligonucleotide (Wang et al 1996). But the mutations are not at specific places, they are dispersed throughout the DNA in the vicinity of the triple helix site.

Agrawal: Did you observe any inhibition of replication in chronically infected cells after cross-linking?

Hélène: Yes. If you cross-link DNA in a plasmid vector you inhibit replication, but if you do not cross-link the oligonucleotide you do not see any effect on replication. If a stable triple helix is formed with an oligophosphoroamidate that is able to arrest transcription, replication is not inhibited, but we have only tried unsubstituted phosphoroamidates, so we will try the same experiment with a covalently attached intercalating agent to determine whether this has any effect on replication.

Stein: I have a question about the oligonucleotide that you used with the sequence of six consecutive guanosines at the 3' end. We've worked with a similar oligonucleotide which doesn't have the stretch containing the cytidine in the midst of a string of thymines. We found that under standard conditions it was almost exclusively a tetraplex. How did you deal with this?

Hélène: There are several ways to avoid tetraplex formation, e.g. by using lithium instead of potassium or by heating and cooling the oligonucleotide rapidly. Obviously, we have the same problems as everybody else, i.e. that the results differ depending on the conditions, and the length of time and temperature at which the oligonucleotides are kept. We are now working with stretches of guanosines that contain a 7-deaza-2'-deoxyguanosine residue because a single 7-deaza-2'-deoxyguanosine residue in a G_6 sequence is sufficient to inhibit quadruplex formation.

Cohen: Does the triplex-binding protein have a natural function, and if so what implications does it have for your strategy?

Hélène: I can only speculate on this. There are several DNA structures *in vivo* that are due to folding of one or both strands. If you have a mirror-repeat sequence you can form an H-DNA triple helix structure that could be recognized by a specific protein. We have no evidence for that. We know that the protein we have identified binds to triple-helical DNA, but that might not be its physiological function.

Inouye: Have you studied both cell extracts and nuclear extracts?

Hélène: Yes, and we obtained the same results in both cases. However, the nuclear extracts could have been contaminated with cytoplasmic proteins and vice versa. Therefore, we do not yet know the localization of this protein.

Inouye: Have you affinity purified this protein?

Hélène: Yes, but we have not yet characterized it.

Toulmé: I understood that you ascribed the inhibition of transcription by the oligonucleotide–intercalator conjugate to local distortion of the template induced by the intercalation rather than by the stabilization of the complex. In this case, wouldn't you expect to observe a difference, depending on the site at which the intercalating agent is linked to the oligonucleotide.

Hélène: Inhibition takes place whether the intercalating agent is on the side where RNA polymerase hits the complex or on the other side. Therefore, it's not solely the distortion that is induced at the intercalation site which is responsible for the inhibitory effect. From what we know of the dissociation rate constants it is difficult to explain the difference between the different oligonucleotides we have tested just on the basis of the lifetime of the complex. There are other factors that we are presently investigating.

Monia: Have you looked at peptide nucleic acid (PNA) analogues for comparison?

Hélène: PNAs induce strand displacement, with a triplex forming on one strand and involving two PNAs and a loop on the other strand. We have not compared PNAs to triple helix-forming oligonucleotides.

Leblen: We have just started a series of experiments with Peter Nielsen in this regard. We know that PNA is able to block transcription elongation by purified RNA polymerases in a cell-free system. In intact cells it is more difficult to assess whether the observed phenotypic response is due to strand displacement by PNAs. PNA does not enter the cell by itself, so we transfected a pre-formed PNA–plasmid DNA target into cells, and it looks as though transcription elongation is blocked.

Hélène: There are many differences between prokaryotic and eukaryotic systems. We also used T3 and T7 polymerases with oligophosphoramidates; they exhibit less inhibition than in a cell-free eukaryotic transcription system.

Leblen: But your transfections were carried out in eukaryotic cells.

Hélène: Yes. Mostly in lymphocytic cell lines.

Matteucci: The formation of triple helices is notorious for having a slow on-rate when the third strand is composed of phosphodiester linkages. Have you looked at the kinetics of phosphoroamidate triple helix formation relative to that for phosphodiesters?

Hélène: Yes. We have compared the phosphodiesters and the phosphoroamidates for the CT oligonucleotides under pH conditions in which the oligophosphodiester binds, i.e. pH 6.5, and we found that phosphoroamidates bind more rapidly than phosphodiesters. We found that there is a hysteresis in the melting curves due to the slow rate of binding of the CT oligophosphodiesters to double-helical DNA, and therefore the melting and cooling curves are different (Rougée et al 1992). However, when we do this with phosphoroamidates the dissociation and association curves are superimposed.

Matteucci: Are the off-rates slower for phosphoroamidates?

Hélène: We don't have all the thermodynamic and kinetic data comparing phosphoroamidates and phosphodiesters. My guess is that the difference in binding free energy cannot be explained simply on the basis of the on-rate. The off-rate is also decreased. We have shown that the dissociation of the triplex is slower with phosphoroamidates than with phosphodiesters.

Monia: I was surprised by the activity of the phosphodiester DNA in the permeabilized cells. I would have thought that it would have been degraded by nucleases. Do psoralens convey any nuclease stability?

Hélène: Psoralen is attached to the 5′ end so it should not prevent 3′ degradation. However, the 3′ end was blocked by an amine group. When we transfected the plasmid with a cross-linked oligonucleotide, even if the 3′end of the oligonucleotide was not protected, when we re-extracted the plasmid the oligonucleotide was still intact. We showed this by hybridization of the oligonucleotide to a complementary probe. Therefore, it seems that once the third strand is bound to the DNA it is protected against degradation, provided that it remains in a triple-helical structure. However, this result may depend on the cell type and the experimental conditions.

References

Degols G, Clarenc J-P, Lebleu B, Léonetti J-P 1994 Reversible inhibition of gene expression by a psoralen-functionalized, triple helix-forming oligonucleotide in intact cells. J Biol Chem 269:16933–16937

Ding D, Gryaznov SM, Lloyd DH et al 1996 An oligodeoxyribonucleotide N3′ → P5′ phosphoramidate duplex forms an α-type helix in solution. Nucleic Acids Res 24:354–360

Musso M, Wang JC, Van Dyke MW 1996 *In vivo* persistence of DNA triple helices containing psoralen-conjugated oligodeoxyribonucleotides. Nucleic Acids Res 24:4924–4932

Rougée M, Faucon B, Mergny JL et al 1992 Kinetics and thermodynamics of triple-helix formation: effects of ionic strength and mismatches. Biochemistry 31:9269–9278

Wang G, Seidman MM, Glazer PM 1996 Mutagenesis in mammalian cells induced by triple helix formation and transcriptional-coupled repair. Science 271:802–805

First- and second-generation antisense oligonucleotide inhibitors targeted against human c-*raf* kinase

Brett P. Monia

Isis Pharmaceuticals Inc., Department of Molecular Pharmacology, 2292 Faraday Avenue, Carlsbad, CA 92008, USA

Abstract. Following extensive screening of more than 50 antisense-designed phosphorothioate oligodeoxynucleotides targeted to human c-*raf* mRNA, one oligodeoxynucleotide (ISIS 5132/CGP 69846A) was identified as being the most potent inhibitor of c-*raf* gene expression both *in vitro* and *in vivo*. ISIS 5132 is a highly sequence-specific and target-specific inhibitor of c-*raf* mRNA and protein levels. c-*raf* inhibition results in dramatic alteration of downstream signalling events within the MAP kinase signalling pathway. Moreover, this oligodeoxynucleotide displays potent antitumour activity against a broad spectrum of tumour types in mouse models and has progressed to Phase I clinical trails. In an effort to identify potential back-up compounds to ISIS 5132, a variety of second-generation 2' sugar modifications have been evaluated for activity against c-*raf* in cell culture. We have identified a number of second-generation oligonucleotides with improved biophysical characteristics that result in enhanced activity against c-*raf* in cell culture. Activity enhancement was most pronounced for 2'-O-methoxyethyl-modified oligonucleotides and this modification also resulted in significantly improved antitumour activity *in vivo*.

1997 Oligonucleotides as therapeutic agents. Wiley, Chichester (Ciba Foundation Symposium 209) p 107–123

We have focused on discovering and developing antisense inhibitors targeted against various members within the MAP kinase signalling pathway for the treatment of cancer and related disorders. In this chapter I will attempt to summarize some of the progress that we have made in utilizing antisense oligonucleotide inhibitors targeted against one particular member of the MAP kinase signalling pathway: c-*raf* kinase. The summary will begin with the *in vitro* characterization of ISIS 5132 (also known as CGP 69846A), a phosphorothioate oligodeoxynucleotide targeted to the 3' untranslated region (UTR) of human c-*raf* mRNA, followed by a review of the *in vivo* antitumour activity displayed by this oligodeoxynucleotide. Experimental evidence supporting an antisense mechanism for this oligodeoxynucleotide, both *in vitro* and *in vivo*, will be covered extensively. Finally, 2'-methoxyethyl (2'-methoxyethoxy) will be introduced as a second-generation chemistry that possesses many attractive biophysical and pharmacological properties relative to simple phosphorothioate oligodeoxy-nucleotides. Antisense effects of this chemical modification within the ISIS 5132 sequence will be summarized both *in vitro* and *in vivo*.

Characterization of ISIS 5132 *in vitro*

To identify antisense oligodeoxynucleotides capable of inhibiting the expression of human c-*raf* kinase, Monia et al (1996a) designed and tested a series of phosphorothioate oligodeoxynucleotides for the inhibition of c-*raf* mRNA levels by treating tumour cells (human A549 lung carcinoma cells) in culture with oligodeoxynucleotides in the presence of cationic lipids and measuring c-*raf* mRNA levels by Northern blotting. This type of screen takes advantage of the ability of phosphorothioate oligodeoxynucleotides to support RNase H-dependent cleavage mechanisms in cells, but it has the limitation that it will not detect antisense inhibitors that act through non-RNA destabilizing mechanisms (Chiang et al 1991, Crooke 1993). Oligodeoxynucleotides used in this analysis were targeted to various regions of the c-*raf* mRNA including the 5 UTR, the translation initiation AUG, the coding region and the 3' UTR, and they were all 20 bases in length. Other groups have reported the activities of antisense-designed oligodeoxynucleotides targeted to the translation initiation AUG of human c-*raf* kinase (Kasid et al 1989, Kasid et al 1996). Reduction in c-*raf* mRNA levels was observed following treatment with only a small subset of oligodeoxynucleotides (Monia et al 1996a). Furthermore, for those oligodeoxynucleotides that did cause reduced c-*raf* mRNA levels, the degree of activity varied greatly. The most potent antisense inhibitor identified from this screen was ISIS 5132, which targets the 3' UTR of the c-*raf* message. This conclusion is based on extensive dose–response experiments comparing oligodeoxynucleotide activity in a 'head-to-head' manner measuring inhibitory effects on both mRNA and protein target levels. The sequence of ISIS 5132 is shown in Table 1. The ability of only a few antisense oligodeoxynucleotides to display potent inhibitory activity against c-*raf* is most likely explained by the role of RNA structure, which has been shown to have profound effects on oligodeoxynucleotide hybridization efficiency (Uhlenbeck 1972, Freier & Tinoco 1975, Herschlag & Cech 1990, Lima et al 1992), as well as the efficacy of antisense oligodeoxynucleotides (Bacon & Wickstrom 1991, Chiang et al 1991).

One of the most critical parameters that must be investigated in order to prove that the mechanism of action of an oligonucleotide is through antisense is by the analysis of specificity. Studies of this nature are generally approached in two ways: measures of oligodeoxynucleotide sequence specificity and measures of target specificity. We have thoroughly examined the sequence requirements for inhibiting c-*raf* mRNA and protein expression *in vitro* by comparing the effects of ISIS 5132 with a series of 'mismatched oligodeoxynucleotides' containing between one and seven mismatches within the ISIS 5132 sequence under dose–response conditions (Monia et al 1996b).

TABLE 1 Design, binding affinities and activities of matched and mismatched phosphorothioate antisense oligodeoxynucleotides targeted against human c-raf kinase mRNA

ISIS no.	Sequence[a]	T_m $(°C)$[b]	K_d $(mM$ at $37°C)$[b]	Inhibition of c-raf mRNA levels in vitro (IC_{50}, nM)[c]	Per cent inhibition of tumour growth in vivo[d]
5132[e]	TCCCGCCTGTGACATGCATT	59.5	0.27	100	80.6
11691	TCCCGCCTGCGACATGCATT	52.5	4.7	225	68.9
11689	TCCCGCCTGCTACATGCATT	45.9	72	500	54.4
11688	TCCCGCCTACTACATGCATT	40.2	730	1000	21.4
11687	TCCCGGCTACTTCATGCATT	35.3	3000	>1000	9.71
11686	TCCCGCCCACTTCATGCATT	30.5	>5000	>1000	0.00
11685	TCCCGCCCACTTGATGCATT	27.2	>5000	>1000	ND
10353	TCCCGGCACTTGATGCATT	—	—	>1000	0.00

[a]Underlined bases indicate mismatches within the ISIS 5132 antisense oligodeoxynucleotide sequence, which targets RNA sequences within the 3′ untranslated region of human c-raf.

[b]T_m and K_d were determined for the indicated phosphorothioate oligodeoxynucleotides targeted to complementary 20mer RNA. K_d values for binding of oligodeoxynucleotides were calculated using Go37 values derived from analyses of melting curves.

[c]Inhibition of c-raf mRNA levels in vitro was determined by Northern blot analysis 24 h after oligodeoxynucleotide treatment, and IC_{50} values were calculated from dose–response curves (results representative of three independent experiments).

[d]Per cent inhibition of tumour growth in vivo was determined at Day 35 by comparing the average tumour volume in animals receiving oligodeoxynucleotides with tumour volumes in animals not receiving oligodeoxynucleotide (saline control).

[e]Also known as CGP 69846A.

ND, not determined. Data taken from Monia et al (1996b).

Melting temperatures (T_m), together with corresponding dissociation constants (K_d), were determined under cell-free conditions for the duplexes of each of these oligodeoxynucleotides with a 20 mer oligoribonucleotide (RNA) complementary to ISIS 5132 (Table 1). As expected for Watson–Crick-based hybridization, affinity decreased as the number of mismatches contained within the ISIS 5132 sequence increased. No co-operative binding was observed for oligodeoxynucleotides containing more than six mismatches.

The IC_{50} value (inhibitory concentration showing a 50% reduction in protein expression) for ISIS 5132-mediated reduction of c-*raf* mRNA levels in A549 tumour cells in culture is approximately 100 nM (Monia et al 1996a). As shown in Table 1, none of the mismatched oligodeoxynucleotides were as potent as ISIS 5132 in inhibiting c-*raf* mRNA expression. Furthermore, inhibition of c-*raf* mRNA levels gradually diminished as the number of mismatches within the ISIS 5132 sequence increased. Incorporation of a single mismatch resulted in a twofold loss in potency. No activity was observed for oligodeoxynucleotides containing more than four mismatches. These findings are the predicted results if the effects of ISIS 5132 on c-*raf* mRNA expression were occurring through a mechanism based on Watson–Crick hybridization to cellular c-*raf* RNA.

In addition to sequence specificity, we have also thoroughly examined the target specificity of ISIS 5132 (Fig. 1). ISIS 5132 is a potent inhibitor of c-*raf* gene expression but it has no effects on the expression of the structurally and functionally related A-*raf* kinase and B-*raf* kinase isotypes, nor on the expression levels of the housekeeping gene glyceraldehyde-3-phosphate-dehydrogenase (Monia et al 1996a). In separate studies, we have identified antisense inhibitors targeted to human A-*raf* and human B-*raf* employing a similar approach as was described for the identification of ISIS 5132. The most active A-*raf* antisense inhibitor identified was ISIS 9069, which targets the translational stop codon of human A-*raf* mRNA. The most active B-*raf* antisense inhibitor discovered was ISIS 13741, which targets the coding region of human B-*raf* mRNA. Employing studies that are similar to those described above for ISIS 5132, we have found that these oligodeoxynucleotides are also highly sequence specific. As shown in Fig. 1, these oligodeoxynucleotides are also highly isotype specific in that they have no effects on the expression levels of other *raf* isoforms.

Characterization of ISIS 5132 *in vivo*

We have characterized the antitumour properties of ISIS 5132 *in vivo* employing nude mouse tumour xenografts. We have demonstrated that ISIS 5132, formulated in simple saline solution, is a potent antitumour agent against a variety of tumour types when administered intravenously at daily doses in the range of 0.01–10 mg oligodeoxynucleotide per kg mouse (Monia et al 1996a,b, Altmann et al 1996, Geiger et al 1997). The relative antitumour activity of ISIS 5132 against various tumour types is shown in Table 2. Since *in vitro* studies employing antisense oligodeoxynucleotides generally require methods that rely on augmentation and modulation of cellular uptake

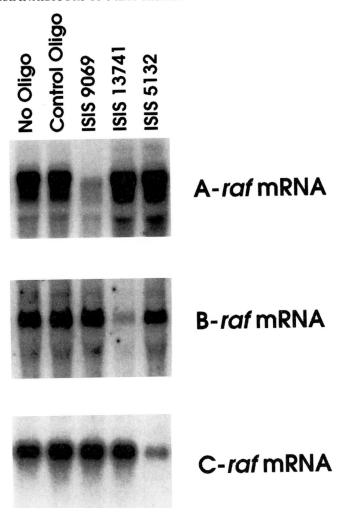

FIG. 1. Isotype-specific inhibition of *raf* expression using antisense oligodeoxynucleotides. Analysis of A-*raf*, B-*raf* and c-*raf* mRNA levels in A549 human tumour cells by Northern blotting following treatment with antisense oligodeoxynucleotides. Control oligo refers to a mismatched oligodeoxynucleotide version of ISIS 5132 (ISIS 10353 shown in Table 1); ISIS 9069 targets the 3′ upstream untranslated region (UTR) of A-*raf* mRNA; ISIS 13741 targets the coding region of B-*raf* mRNA; ISIS 5132 targets the 3′ UTR of c-*raf* mRNA.

(Bennett et al 1992), significant efforts must be undertaken to demonstrate that the pharmacological effects observed *in vivo* are related to the effects observed *in vitro*, and that they are based on an antisense mechanism of action. A number of studies have been reported demonstrating antitumour effects of antisense oligodeoxynucleotides *in vivo* that are consistent with an antisense mode of action (Gray et al 1993, Schwab et al

1994, Bennett et al 1995, Nesterova & Cho-Chung 1995, Monia et al 1996a,b, Dean et al 1996, Yazaki et al 1996, Altmann et al 1996, Geiger et al 1997), some of which involve targeting members of the MAP kinase signalling pathway (Schwab et al 1993, Bennett et al 1995, Monia et al 1996a,b). Evidence supporting an antisense mechanism *in vivo* can be obtained in a number of ways, and this subject has been the topic of numerous reviews (Wagner 1994, Stein & Kreig 1994, Crooke 1996).

The primary support for an antisense mechanism of action *in vivo* should come from at least one of two different experimental approaches: (1) the demonstration of reduced target RNA or protein expression; and/or (2) the demonstration of oligonucleotide sequence specificity for the pharmacological effects observed for the antisense compound. Stating the second approach more simply, a group of 'control' oligonucleotides should not, at the same dose, induce the same effects as the antisense compounds, and the rank order potencies of mismatched analogues should be predicted from Watson–Crick hybridization mechanisms. We have succeeded in accomplishing both of these objectives in addressing the mechanism of *in vivo*

TABLE 2   Spectrum of *in vivo* antitumour activity displayed by ISIS 5132 employing nude mouse tumour xenografts

Tumour type[a]	Relative antitumour activity[b]
A549 (lung)	+++++
T24 (bladder)	+++++
MDA-MB-231 (breast)	+++++
PC-3 (prostate)	++++
SK-MEL-1 (melanoma)	++++
WiDr (colon)	+++
HCT116 (colon)	+++
NCI-H209 (small-cell lung)	+++
NCI-H520 (lung)	+++
MCF7 (breast)	+++
SK-MEL-3 (melanoma)	+++
DU 145 (prostate)	++
COLO 205 (colon)	++
NCI-H69 (small-cell lung)	+
NCI-H460 (lung)	+
NIH:OVCAR-3 (ovary)	Not active

[a]For all (human) tumour types, animals received daily intravenous doses of oligodeoxynucleotide (in saline) in the range of 0.06–6.0 mg/kg.
[b]A relative antitumour scale is shown in which the number of '+' symbols correlates with the efficacy of ISIS 5132 against that particular tumour type.
See Muller et al (1997) for experimental details.

antitumour activity of ISIS 5132 in nude mouse tumour xenograft models (Monia et al 1996a,b). We have demonstrated that the administration of ISIS 5132 to mice bearing subcutaneously implanted A549 tumours results in a time-dependent reduction in c-*raf* mRNA levels in tumours as determined by Northern blot analysis (Monia 1996a). However, no effects on c-*raf* mRNA levels were observed following administration of a control oligodeoxynucleotide. We have also demonstrated that the antitumour effects of ISIS 5132 are highly sequence specific and that rank order antitumour potency of mismatched analogues is predicted from Watson–Crick hybridization mechanisms (Monia 1996b). As shown in Table 1, the antitumour effects of ISIS 5132 and mismatched oligodeoxynucleotides containing between one and five mismatches correlates well with the affinity of these oligodeoxynucleotides for RNA containing the complementary sequence of ISIS 5132 under cell-free conditions and with the effects on c-*raf* mRNA expression observed *in vitro*. Even the incorporation of a single mismatch resulted in a significant loss of antitumour activity. No antitumour activity was observed for oligodeoxynucleotides containing more than two mismatches. These results, together with the effects observed on c-*raf* mRNA levels in tumours *in vivo*, strongly support an antisense mechanism underlying the *in vivo* antitumour properties of ISIS 5132.

2'-Methoxyethyl as a second-generation chemistry

ISIS 5132 is a potent antitumour agent that displays low toxicity in animals. Thus, this compound displays a large therapeutic index in preclinical animal models, relative to most classical antitumour agents, and has progressed rapidly to clinical trials against solid tumours. In an ongoing effort to improve the pharmacological and toxicological properties of this and other antisense compounds, focus has been placed on the 2' sugar position of nucleosides, and this topic has been covered extensively in numerous reviews (Uhlmann & Peyman 1990, Milligan et al 1993, Sanghvi & Cook 1993, De Mesmaeker et al 1995, Altmann et al 1996). One of our goals was to determine whether it would be possible to reduce phosphorothioate content in oligonucleotides by the incorporation of certain 2' sugar modifications whilst maintaining or improving pharmacological activity. Since phosphorothioate linkages have been reported to exert some undesired pharmacological effects due to non-specific protein binding, this approach theoretically could result in oligonucleotides with greater specificity and less toxicity. Since phosphorothioate modifications are responsible for the nuclease stability of ISIS 5132 in biological systems, one of the greatest objectives for this approach was to identify a 2' sugar modification that conferred sufficient stability towards nucleases. Furthermore, since ISIS 5132 acts through an RNase H-dependent mechanism and 2' sugar modified oligonucleotides are not substrates for RNase H (Monia et al 1993, 1996a), a chimeric oligonucleotide approach was employed that utilizes a stretch of eight DNA-phosphorothioate nucleotides that support RNase H activity flanked by 2' sugar-

$T_oC_oC_oC_oG_oC$ $_sC_sT_sG_sT_sG_sA_sC_sA_s$ $T_oG_oC_oA_oT_oT$
2' Sugar Modified   2' Sugar Modified

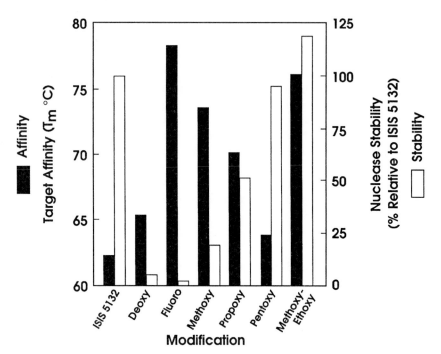

FIG. 2.   Biophysical properties of 2'-sugar-modified chimeric oligonucleotide derivatives of ISIS 5132. The sequence of oligonucleotide ISIS 5132 is shown at the top with nucleotides that contain 2'-sugar modifications indicated: phosphodiester linkages are indicated with a subscript 'o'; and phosphorothioate linkages are indicated with a subscript 's'. Melting temperatures (target affinity) were determined as described by Monia et al (1996b). Relative nuclease stability was determined against snake venom phosphodiesterase as described previously in Monia et al (1996c) and Altmann et al (1996). Quantitation was also as described previously (Monia et al 1996c).

modified phosphodiester 'wings' (Fig. 2) (Altmann et al 1996). Thus, this type of molecule contains both a mixed 2' sugar composition and a mixed backbone structure.

Figure 2 shows the effects on target affinity and nuclease stability under cell-free conditions of various 2' sugar modifications incorporated into the ISIS 5132 chimeric oligonucleotide. Affinity was determined by measuring the T_m of each 2' sugar-modified chimera against a complementary RNA (Monia et al 1993). Nuclease stability was determined by measuring the rate of oligonucleotide degradation in the presence of snake venom phosphodiesterase (Monia et al 1996c, Altmann et al 1996).

The modification that conferred the greatest improvement in affinity for RNA was the 2'-fluoro-modified chimera. However, this compound displayed a level of resistance to nuclease degradation that was substantially less than that of the phosphorothioate DNA molecule, indicating that the usefulness of this chemistry in a phosphodiester backbone is of limited value. A similar result was observed for the 2'-methoxy modified chimera. The modification that conferred the best combination of improved affinity and nuclease stability was the 2'-methoxyethyl (or 2'-methoxyethoxy) modification. These results suggest that the biophysical properties of this modification may be sufficient to allow for it to be an effective antisense inhibitor when incorporated into appropriate antisense molecules.

We have tested the ISIS 5132 methoxyethyl mixed backbone chimera for antisense activity in cell culture and for antitumour activity *in vivo* and compared it with the effects of ISIS 5132 (Altmann et al 1996). The 2'-methoxyethyl chimera was found to be a more potent inhibitor of c-*raf* mRNA levels in cell culture relative to ISIS 5132 (Fig. 3). This effect is most likely explained by the enhanced stability and affinity conferred by this modification. The activity of other 2' sugar modifications against c-*raf* mRNA levels was also measured and found to correlate well with the relative resistance of these molecules to nuclease degradation under cell-free conditions. The 2'-propoxy mixed backbone chimera, which displayed modest resistance to degradation, showed modest activity against c-*raf* kinase whereas a 2'-methoxy mixed backbone chimera, which displayed no resistance to nuclease degradation, was completely inactive as an antisense inhibitor.

To determine whether the 2'-methoxyethyl mixed backbone chimera would be effective as an antitumour agent in animals, we tested this compound for antitumour activity against A549 tumour xenografts under dose–response conditions and compared its activity to that of ISIS 5132 (Fig. 4). The 2'-methoxyethyl chimera was found to be a potent antitumour agent in this model, displaying activity slightly but significantly better than that of the simple phosphorothioate ISIS 5132. In other studies, we have measured the levels of both ISIS 5132 and the methoxyethyl chimera in animal tissues and in tumours, and we have detected both compounds intact together with their predicted metabolites. These results indicate that the 2'-methoxyethyl modification is an effective antisense chemistry when incorporated into phosphodiester oligonucleotides for both *in vitro* and *in vivo* purposes.

Conclusions

These studies demonstrate that antisense inhibitors can be employed successfully for abrogating the function of the MAP kinase signalling pathway *in vitro* and *in vivo* by targeting c-*raf* kinase, and that this pathway is an attractive target for the discovery of novel anticancer agents. Furthermore, these studies provide strong evidence that the pharmacological activity observed for ISIS 5132 both *in vitro* and *in vivo* is occurring through an antisense mechanism of action and that antisense is a valuable approach for both target validation and for the discovery of novel therapeutic agents

$T_OC_OC_OC_OG_OC$ $_sC_sT_sG_sT_sG_sA_sC_sA_s$ $T_OG_OC_OA_OT_OT$

2' Sugar Modified   2' Sugar Modified

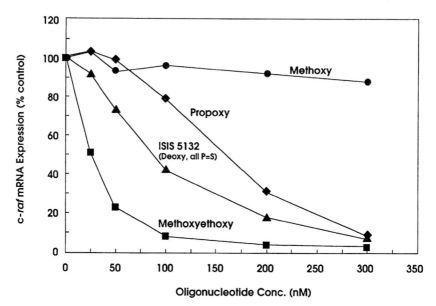

FIG. 3. Inhibition of c-*raf* mRNA levels by 2'-sugar-modified chimeric antisense oligonucleotides in cultured T24 human bladder carcinoma cells. c-*raf* mRNA levels were determined by Northern blot analysis in T24 cells treated with antisense oligonucleotides at the indicated concentrations. The 2'-sugar modifications—2'-methoxy, 2'-propoxy and 2'-methoxyethoxy—were compared to the activity of the 'parent' phosphorothioate oligodeoxynucleotide ISIS 5132. The sequence of ISIS 5132 is shown at the top with nucleotides that contain 2'-sugar modifications indicated: phosphodiester linkages are indicated with a subscript 'o'; and phosphorothioate linkages are indicated with a subscript 's'. Methods for oligonucleotide treatment, RNA analysis, normalization and quantitation were as described by Monia et al (1996c). P=S, phosphorothioate.

for the treatment of human disease. First-generation antisense chemistries (phosphorothioate oligodeoxynucleotides) are extremely valuable and attractive pharmacological agents, and many have progressed to clinical trials. Second-generation chemistries (e.g. 2'-methoxyethyl) appear to have improved pharmacological and toxicological properties in preclinical models and should be in the clinic in the near future. Further improvements in oligonucleotide chemistry ('third generation' and beyond) will be a continuing process, aimed at improving therapeutic index further by increasing potency, decreasing toxicity or addressing other pharmaceutical aspects such as oral bioavailability. Thus, although tremendous progress has been made in antisense technology both from a biological and a chemical

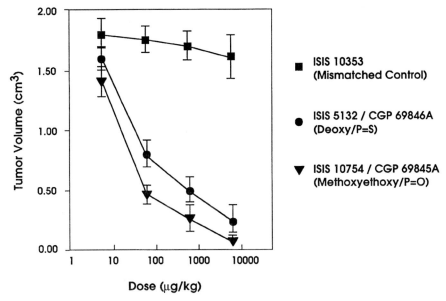

FIG. 4. Antitumour activity of antisense oligonucleotide targeted to human c-*raf* mRNA. ISIS 5132 and ISIS 10754 (2′-methoxyethoxy mixed backbone chimeric analogue of ISIS 5132, also known as CGP 69845A) were formulated in saline and administered to nude mice bearing subcutaneously implanted tumours (A549 human adenocarcinoma) by daily intravenous injection under dose–response conditions. Tumour volume was determined 27 days following implantation of tumour fragments and 17 days following initiation of oligonucleotide treatment as described previously in Altmann et al (1996). Oligonucleotide treatment was initiated after the tumour reached a mean volume of 100 mm³. Also shown is the lack of antitumour activity exhibited by a mismatched phosphorothioate oligodeoxynucleotide analogue of ISIS 5132. P=O, phosphodiester; P=S, phosphorothioate.

standpoint, we are just beginning to understand and exploit this technology to serve both therapeutics as well as basic research.

Acknowledgements

I thank a number of research colleagues who have contributed substantially to the findings and conclusions described in this chapter: Doriano Fabbro, Marcel Muller, Thomas Geiger, Karl-Heinz Altmann, Robert Haner, Heinz Moser, Paul Nicklin and Judy Phillips of Novartis Pharmaceuticals (Switzerland); and P. Dan Cook, Kathy McGraw, Edward Wancewicz and Joseph Johnston of Isis Pharmaceuticals.

References

Altmann K-H, Dean N M, Fabbro D et al 1996 Second generation of antisense oligonucleotides: from nuclease resistance to biological efficacy in animals. Chimia 50:168–176

Bacon TA, Wickstrom E 1991 Walking along human c-*myc* mRNA with antisense oligodeoxynucleotides: maximum efficacy at the 5' cap region. Oncogene Res 6:13–19

Bennett CF, Chiang H, Shoemaker JE, Mirabelli CK 1992 Cationic lipids enhance cellular uptake and activity of phosphorothioate antisense oligonucleotides. Mol Pharmacol 41:1023–1033

Bennett CF, Dean N, Ecker DJ, Monia BP 1995 Pharmacological activity of phosphorothioate antisense oligonucleotides: *in vivo* studies. In: Agrawal S (ed) Methods in molecular medicine: antisense therapeutics. Humana Press, New York, p 13–46

Chiang MY, Chen H, Zounes MA, Freier SM, Lima WF, Bennett CF 1991 Antisense oligonucleotides inhibit intercellular adhesion molecule 1 expression by two distinct mechanisms. J Biol Chem 266:18162–18172

Crooke ST 1993 Therapeutic applications of oligonucleotides. Annu Rev Pharmacol Toxicol 32:329–376

Crooke ST 1996 Proof of mechanism of antisense drugs. Antisense Nucleic Acid Drug Dev 6:145–147

Dean NM, McKay R, Miraglia L 1996 Inhibition of growth of human tumor cell lines in nude mice by an antisense oligonucleotide inhibitor of protein kinase C alpha expression. Cancer Res 56:3499–3507

De Mesmaeker A, Häner R, Martin P, Moser HE 1995 Antisense oligonucleotides. Acc Chem Res 28:366–374

Freier SM, Tinoco I 1975 The binding of complementary oligoribonucleotides to yeast initiator transfer RNA. Biochemistry 14:3310–3314

Geiger T, Müller M, Monia BP, Fabbro D 1997 Antitumor activity of a C-raf antisense oligonucleotide in combination with standard chemotherapeutic agents against various human tumors transplanted subcutaneously into nude mice. Clin Cancer Res 3:1179–1185

Gray GD, Hernandez OM, Hebel D, Root M, Pow-Sang JM, Wickstrom E 1993 Antisense DNA inhibition of tumor growth by c-Ha-*ras* oncogene in nude mice. Cancer Res 53:577–580

Herschlag D, Cech TR 1990 Catalysis of RNA cleavage by the *Tetrahymena thermophila* ribozyme: kinetic description of the reaction of an RNA substrate that forms a mismatch at the active site. Biochemistry 29:10172–10780

Kasid U, Pfeifer A, Brennan, T et al 1989 Effect of antisense c-*raf-1* on tumorigenicity and radiation sensitivity of a human squamous carcinoma. Science 243:1354–1356

Kasid U, Simeng S, Dent P, Ray S, Whiteside TL, Sturgill TW 1996 Activation of *raf* by ionizing radiation. Nature 382:813–816

Lima W, Monia BP, Freier SM, Ecker DJ 1992 Implication of RNA structure on antisense oligonucleotide hybridization kinetics. Biochemistry 31:12055–12061

Milligan JF, Matteucci MD, Martin JC 1993 Current concepts in antisense drug design. J Med Chem 36:1923–1926

Monia BP, Lesnik E, Gonzalez C et al 1993 Evaluation of 2'-modified oligonucleotides containing deoxy gaps as antisense inhibitors of gene expression. J Biol Chem 268:14514–14522

Monia BP, Johnston JF, Geiger T, Muller M, Fabbro D 1996a Antitumor activity of a phosphorothioate antisense oligodeoxynucleotide targeted against c-*raf* kinase. Nat Med 2:668–672

Monia BP, Sasmor H, Johnston JF et al 1996b Sequence-specific antitumor activity of a phosphorothioate oligodeoxyribonucleotide targeted to human c-*raf* kinase supports an antisense mechanism of action *in vivo*. Proc Natl Acad Sci USA 93:15481–15484

Monia BP, Johnston JF, Sasmor H, Cummins L 1996c Nuclease resistance and antisense activity of modified oligonucleotides targeted to Ha-*ras*. J Biol Chem 14533–14540

Nesterova M, Cho-Chung Y 1995 A single-injection protein kinase A-directed antisense treatment to inhibit tumor growth. Nat Med 1:528–533

Sanghvi YS, Cook PD 1993 Towards second-generation synthetic backbones for antisense oligonucleotides. In: Chu CK, Beker DC (eds) Nucleosides and nucleotides as antitumor and antiviral agents. Plenum Press, New York, p 311–319

Schwab G, Chavany C, Duroux I 1994 Antisense oligonucleotides absorbed to polyalkylcyanoacrylate nanoparticles specifically inhibit mutated H-ras-mediated cell proliferation and tumorgenicity in nude mice. Proc Natl Acad Sci USA 91:10460–10464

Stein C, Krieg AM 1994 Problems in interpretation of data derived from in vitro and in vivo use of antisense oligodeoxynucleotides. Antisense Res Dev 4:67–69

Uhlenbeck OC 1972 Complementary oligonucleotide binding to transfer RNA. J Mol Biol 65:25–41

Uhlmann E, Peymann A 1990 Antisense oligonucleotides: a new therapeutic principle. Chem Rev 90:543–545

Wagner RW 1994 Gene inhibition using antisense oligodeoxynucleotides. Nature 372:333–335

Yazaki T, Ahmed S, Chahlavi A 1996 Treatment of glioblastoma 0–87 by systemic administration of an antisense protein kinase C—a phosphorothioate oligodeoxynucleotide. Mol Pharmacol 50:242–263

DISCUSSION

Inouye: You mentioned that the levels of mRNA were reduced. Could you clarify whether the production of mRNA was reduced or the stability of mRNA?

Monia: We measure steady-state levels of mRNA by Northern blotting. We knockout all the mRNA in the cell by reducing its stability through an RNase H mechanism, i.e. the oligonucleotide binds to mRNA, which causes the recruitment of RNase H and the subsequent cleaving of mRNA. We do not affect transcription of new mRNA.

Eckstein: In your in vivo experiment you showed that there was a decrease in mRNA over a 13-day period, and I presume that this correlates with the decrease in tumour load, although you didn't show this. You then showed that the tumour gradually returns, but does this correlate with the levels of mRNA?

Monia: In that particular tumour cell line (the A549 human lung carcinoma cell line) we saw a slight regrowth of the tumour at later time points, but that did not correlate with the mRNA levels. This suggests that the antisense oligonucleotide is still working. Bear in mind that regrowth of tumours is a common observation with low molecular weight anticancer agents in xenograft models, and therefore this might represent an artefact of this type of model. However, it is possible that compensatory mechanisms involving other signalling pathways are activated following inhibition of certain signalling molecules and that the emergence of a 'resistant population' is explained by compensatory mechanisms involving other signalling pathways. The emergence of a 'resistant population' can, therefore, be explained by the ability of the tumours to recruit other signalling pathways, although we haven't demonstrated that this occurs.

Gewirtz: Were the animal data that you presented relevant to established tumours? In other words, did you inoculate the tumour, administer the oligonucleotide and look at growth over time?

Monia: We allowed the tumour to grow to a certain size before we initiated treatment with the oligonucleotide, and then we measured tumour volume over time.

Gewirtz: What effects are observed in more established tumours?

Monia: We have looked at the effects of administering the protein kinase C (PKC)-α oligonucleotide in a glioblastoma tumour model, in which we changed treatments between the antisense oligonucleotide and the control oligonucleotide. Half way through the study, when the tumours were fairly large in the control treatment, we switched regimens and showed that the PKC-α oligonucleotide had a potent antitumour effect.

Gewirtz: It prevents further growth but did you observe regression of the tumour?

Monia: In that study we did not observe regression. In other tumours we have observed tumour regression with certain oligonucleotides. It depends on the tumour type.

Inouye: Can you explain why there are differences in sensitivity?

Monia: RNA structure is probably playing the principle role in determining which oligonucleotides are effective and which are not. Certainly, other things can play a role, as well as the ability of RNase H to recognize the duplex efficiently and cause cleavage. However, in my opinion, the principle explanation is simply the accessibility of an antisense oligonucleotide to RNA based on RNA structure.

Inouye: Do the antisense oligonucleotides bind to several regions?

Monia: Yes. However, although it looks like there's a gradual effect as you approach the binding site, the oligonucleotides are spaced out with fairly wide gaps in-between. We have about 10 or 12 oligonucleotides targeted to the c-*raf* 3′ untranslated region (UTR), which are all 20 mers, but the 3′ UTR is about 3000 bp, so we've only covered a small region.

Inouye: Can the antisense targets be predicted from these results?

Monia: No. The method is purely empirical. There are certain hot spots in an RNA that tend to be good places for antisense oligonucleotides to bind and be effective but this is not universally true for all targets. Three of our best oligonucleotides for different targets happen to be targeting the 3′ UTR but that's probably a coincidence.

Toulmé: How did you optimize the gap size? Was it related to the affinity for the target?

Monia: We used a cellular assay for activity. We designed oligonucleotides with increasing gap sizes in various places within the 20 mer antisense molecule, and we measured activity by dose–response comparisons determining mRNA levels. We thoroughly examined the gap size and found that it affects the effectiveness of the molecule. Chimeric oligonucleotide design must be optimized for each sequence.

Hélène: Do you observe inhibition of tumour growth with A-*raf* and B-*raf* kinase antisense oligonucleotides?

Monia: We have only looked at A-*raf* because B-*raf* inhibitors have been identified only recently. We have seen some antitumour activity with the A-*raf* antisense oligonucleotide but it's much less potent than the c-*raf* antisense inhibitor. However, we have been unable to identify a phosphorothioate oligonucleotide against A-*raf* that is as potent as c-*raf*, which can explain the antitumour data in one of two ways: either A-*raf* isn't the best target compared to c-*raf*; or the inhibitor is better for c-*raf*.

Hélène: In many cases if one isotype is inhibited, one of the others can take over.

Monia: Yes. We would also like to study them in combination.

Lebleu: Is there any correlation between the susceptibilities of various tumours to these oligonucleotides and the ability to induce apoptosis?

Monia: We don't have a molecular explanation as to why certain tumours are more sensitive than others. c-*raf* is being expressed in those tumours but there is no correlation between the expression levels and the sensitivity. The apoptosis studies were done on only one cell line, so it is too soon to determine whether there is a correlation between apoptosis and sensitivity.

Agrawal: You showed that your oligonucleotide had variable activity *in vivo*, i.e. for certain tumours it was effective and for others it was not. How do you explain the activity of the oligonucleotides in cell lines derived from these tumours?

Monia: All of the tumour cell lines are sensitive to c-*raf* inhibition *in vitro* when we use cationic lipids, despite the fact that there are clear differences in the mRNA and protein levels for c-*raf*. So, differential sensitivity *in vivo* can't simply be explained by the molecule not working in tumour cells *in vitro*.

Akhtar: Does the distribution of oligonucleotides vary in different tumours?

Monia: We haven't looked but we wouldn't expect there to be any variation.

Southern: Is it possible that A-*raf* and B-*raf* are expressed in the resistant tumour types, but not in the sensitive ones, indicating that there is redundancy in the resistance?

Monia: It is possible and therefore worth investigating. We do know that A-*raf* and c-*raf* are expressed ubiquitously, whereas B-*raf* expression tends to be more cell-type specific.

Iversen: How many of the tumours sensitive to c-*raf* also have a point mutation in *ras*? In other words, is the defect in the transformed cell upstream or downstream of *raf*?

Monia: We have only looked for *ras* mutations in a few tumours. However, some of the more sensitive ones, such as the T24 human bladder carcinoma cell line, clearly have *ras* mutations. On the other hand, although it hasn't been typed, breast tumours rarely contain *ras* mutations, and we have a breast tumour cell line that is extremely sensitive to the c-*raf* antisense oligonucleotide. Also, certain tumours that have *ras* mutations are only moderately sensitive to the oligonucleotide. Therefore, the situation is much more complicated than simply saying *ras* mutations make tumours sensitive to c-*raf* antisense oligonucleotides.

Crooke: In these tumour studies we're seeing activity at doses that are about 100th of the maximum tolerated dose (MTD). One of the complexities of interpreting these studies, compared to traditional cytotoxic studies, is that all of the traditional cytotoxic data are done at the MTD, and the MTD in these studies is probably around 200 mg/kg. Until we do more high dose experiments it's risky to draw too many conclusions about the spectrum of activity.

Iversen: Have you looked at the same spectrum using a combination of PKC inhibitor and c-*raf* inhibitor?

Monia: We've looked at this combination in one tumour cell line and we've seen a slightly additive antitumour effect, but nothing that would suggest synergy.

Crooke: One of our objectives is to develop an anticancer drug with a decent therapeutic index. There are only a couple of drugs that have a therapeutic index greater than one. One of the basic premises is that in a variety of diseases an isotype of a multigene family will tend to dominate. Therefore, multiple isotype inhibitor combination studies are interesting pharmacologically and will teach us a lot. However, from a therapeutic point of view I'm disinclined to do those studies because we will lose what we have already gained, i.e. isotype selectivity.

Wickstrom: How similar is the ISIS 5132 equivalent target sequence in mice compared to the human gene? Because this has implications in terms of the potential for toxicity in humans.

Crooke: For every drug that we take to the clinic, we test the human compound, so that we have a clear understanding of the chemical toxicity, and then we create the mouse, monkey or rat version (whichever has greatest homology). The sequences are usually different but most of the time we can create an oligonucleotide that works roughly in the same region in the gene.

Hélène: In the studies you performed with Brett Monia, where you demonstrated synergy between a cytotoxic agent and antisense oligonucleotides, did you look at cell lines that are resistant to the cytotoxin as well as those that are sensitive? If you select cytotoxin-resistant cells in a population do you still observe cytotoxin resistance when you inhibit *raf* kinase?

Monia: We have only looked at two cisplatin-resistant tumours. When we tested the ISIS 5132 antisense oligonucleotide it showed antitumour activity, but a combination with cisplatin was much better than using ISIS 5132 alone. One explanation is that the Raf signalling cascade is also involved in mediating cisplatin resistance in that particular tumour.

Crooke: The tumours were originally insensitive to cisplatin, i.e. they were cisplatin-resistant tumours that were made resistant to cisplatin. You're addressing two questions: one involving primary sensitivity; and the second involving mechanisms of resistance—i.e. multi-resistance protein (MRP) and multiple drug resistance (MDR) protein etc.—of which we don't have any real data, although we do have inhibitors of MRP and MDR protein so we can look at the effects of those in resistant populations.

Gait: Cisplatin is known to react with the N7 positions of guanosine. Is there any evidence that cisplatin is getting inside the cell via complexation with the oligonucleotide?

Monia: There's no evidence, but we haven't looked at that specifically. Cisplatin was only administered once a week because of its toxic effects, whereas the oligonucleotides were administered daily. I would expect most of the cisplatin to be washed out for most of the study.

Gewirtz: Are there any changes in oligonucleotide uptake in the drug-treated models?

Monia: We haven't looked. Paul Nicklin has looked at the effects of oligonucleotide uptake in tumours, but we haven't looked at whether cytotoxins augment the uptake of oligonucleotide and vice versa.

Crooke: There is no evidence to suggest that the uptake of oligonucleotides in tumours is enhanced by cisplatin.

Differential oligonucleotide activity in cell culture versus mouse models

Eric Wickstrom and Frederick L. Tyson*

*Department of Microbiology and Immunology and Kimmel Cancer Center, Thomas Jefferson University, 1025 Walnut Street, Philadelphia, PA 19107, and *National Institute of Environmental Health Sciences, Research Triangle Park, NC 27709, USA*

Abstract. The usual course of drug discovery begins with the demonstration of compound activity in cells and, usually, a lower level of activity in animals. Successive rounds of drug design may result in a compound with sufficient activity in animals to justify clinical trials. The basic endpoints of therapeutic oligonucleotide experiments include target antigen reduction, target messenger reduction and inhibition of transformed cell proliferation or viral replication. However, one should expect oligonucleotides to exhibit pleiotropic behaviour, as do all other drugs. In an animal oligonucleotides will necessarily bind to and dissociate from all macromolecules encountered in the blood, in tissues, on cell surfaces and within cellular compartments. Contrary to expectations, oligonucleotides designed to be complementary to certain transcripts have sometimes been found moderately effective in cell-free extracts, more effective in cell culture and most effective in animal models. If greater potency against standard endpoints is reported in mouse models than was observed in cell culture, critical examination must consider alternate modes of action in animals that may not apply in cell culture. This counterintuitive paradox will be examined, based on studies of Ha-*ras* expression in bladder cancer, Ki-*ras* expression in pancreatic cancer, *erb*B2 expression in ovarian cancer and c-*myc* expression in B cell lymphoma.

1997 Oligonucleotides as therapeutic agents. Wiley, Chichester (Ciba Foundation Symposium 209) p 124–141

The concept (Belikova et al 1967) and demonstration (Zamecnik & Stephenson 1978) of complementary oligonucleotides as a method to inhibit expression of pathogenic genes has led to intensive investigation and development of therapeutic oligonucleotides over the past 30 years (Wickstrom 1991, Crooke & Lebleu 1993, Akhtar 1995, Agrawal 1996a, Wickstrom 1997). Unexpectedly, a number of therapeutic oligonucleotides have displayed greater potency in animals than in cell culture.

Is this paradoxical observation the result of several non-specific modes of toxicity, or can one identify naturally occurring vehicles for oligonucleotide distribution and cellular uptake that parallel the role of cationic lipids often used in cell culture? Four examples from the experience of this laboratory are presented which demonstrate the phenomenon. The data do not support the hypothesis of random toxicity by all oligonucleotide sequences contributing to antitumour effects. The natural carrier hypothesis, however, has not been rigorously tested.

Ha-ras-induced bladder cancer

Humans carry three functional Ras proto-oncogenes, Ha-*ras*, Ki-*ras*, and N-*ras*, coding for 21 kDa proteins 188–189 amino acids long (Lowy & Willumsen 1993). Ki-*ras*, Ha-*ras* and N-*ras* have been detected in more human tumour types and at higher frequencies than any other oncogenes (Bishop 1991). *ras* p21 protein, anchored by fatty acid tails to the inner surface of the plasma membrane, binds GTP, creating the active proliferation signal, then hydrolyses GTP to GDP, recreating the inactive form. Human bladder cancer EJ or T24 cells contain a $G \rightarrow U$ mutation in the middle of codon 12 of Ha-*ras* mRNA (Tabin et al 1982), predicted to occur in an accessible hairpin loop of the 1.2 kb message (Daaka & Wickstrom 1990). This mutation is responsible for the $G12 \rightarrow V12$ substitution in the GTPase site of p21, which reduces GTPase activity and thus elevates the baseline level of Ras●GTP. This induces continuous cell proliferation, which permits tumour formation. Phosphodiester oligonucleotides complementary to sites in Ha-*ras* mRNA have been observed to inhibit *ras* p21 protein expression and cellular proliferation in mutant Ha-*ras*-transformed murine cells in a dose- and sequence-dependent manner (Daaka & Wickstrom 1990). The most potent of those sequences was used to pretreat mutant Ha-*ras*-transformed murine cells, which were then implanted subcutaneously into nude mice (Gray et al 1993). Such xenografts yielded large, easily monitored tumours within two weeks, providing a simple initial model for evaluating antisense therapy *in vivo*. Tumour growth was powerfully suppressed, dependent upon the oligonucleotide sequence, for at least 15 days after exposure to complementary DNA against the Ha-*ras* target.

Moving around the 12th codon loop with different complementary DNAs to elucidate the structural basis for these observations, Bennett et al (1996) found strongest binding to the 5′ side of the hairpin loop. These observations implied that the 5′ side of the loop, the site of the activating mutation, was a more favourable antisense target than the 3′ side. The length of anti-Ha-*ras* DNA was then varied from 5 to 25 nt, and a *ras*–luciferase expression construct was used to assay the level of Ha-*ras* expression. DNA phosphorothioates targeted to the 5′ side of the hairpin loop were able to inhibit expression of the *ras*–luciferase fusion protein at concentrations as low as 10 nM, with the aid of lipofectin. RAS17 (5′ dCCACACCGACGGCGCCC 3′) was the shortest oligomer to give maximum inhibition. These results implied that sequence-specific antisense DNAs against Ha-*ras* mRNA might be effective antitumour agents for clinical administration.

In this laboratory, tumorigenesis by human EJ bladder carcinoma cells implanted subcutaneously into nude mice now serves as a more clinically relevant model of

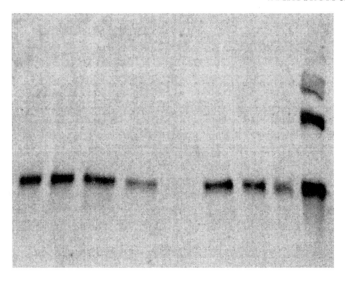

A   B   C   D   E   F G   H   I

FIG. 1.   Inhibition of total EJ human bladder cancer cell *ras* p21 protein levels by DNA phosphorothioates. Following 5 h of exposure to oligonucleotides, with or without Lipofectamine (1 : 6, w/w; Gibco/BRL) in Opti-Mem (Life Technologies, Gaithersburg, MD) at 37 °C, EJ cells were incubated for an additional 48 h in Dulbecco's modified Eagle's medium plus 10% fetal bovine serum. Cell lysates were processed for Western blot analysis. Lane A, no DNA, no lipid; lane B, no DNA but 3 μg/ml lipid; lane C, no DNA but 9 μg/ml lipid; lane D, 0.1 μM RAS17, 3 μg/ml lipid; lane E, 0.3 μM RAS17, 9 μg/ml lipid; lane F, 0.3 μM RAS17 but no lipid; lane G, 0.1 μM scrambled control, 3 μg/ml lipid; lane H, 0.3 μM scrambled control, 9 μg/ml lipid; lane I, p21 standard, 5 ng.

Ha-*ras*-dependent solid tumours. In cell culture we have observed that RAS17 phosphorothioate oligonucleotide transfected with Lipofectamine (Gibco/BRL) suppressed p21 expression effectively at 0.3 μM (Fig. 1), in agreement with previous experiments using RAS17 against a *ras*–luciferase construct (Bennett et al 1996).

A tumour suppression experiment was then performed in nude mice following subcutaneous implantation of 10^6 EJ cells into each of 10 mice in each group, administering 200 nmoles of RAS17, a scrambled control or a phosphate-buffered saline (PBS) vehicle subcutaneously every three days. Under this regimen, serum oligonucleotide concentrations, as measured by OliGreen™ fluorescence (Gray & Wickstrom 1997), declined to 30 μM by 1 h after administration, about three elimination half-lives (DeLong et al 1997). Under these conditions, tumour concentration of phosphorothioate would be about 5% of the whole-body dose (DeLong et al 1997), or about 1.5 μM. By 2 h, after another three half-lives, tumour concentration would be down to 0.2 μM.

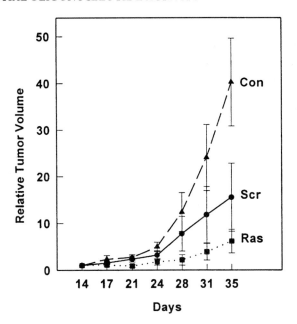

FIG. 2. Inhibition of tumour growth by subcutaneous DNA phosphorothioates. Tumours were induced in female BALB/c *nu/nu* mice by subcutaneous injection of 10^6 EJ human bladder cancer cells on Day 0. Subcutaneous DNA administration began on Day 0 on the flank opposite the site of EJ injection, with a dose of 200 nmol every three days for 31 days, for a total dose of $2\,\mu$mol, or $10\,\mu$g. Tumour volumes were measured for each animal, then normalized to the tumour volume for each animal on Day 14, the first day on which all animals were presented with a palpable tumour. The relative tumour volumes represent mean $\pm$ S.E.M. for groups of 10 mice. Symbols: ▲ Con, phosphate-buffered saline vehicle; ● Scr, scrambled 17 mer; ■ Ras, RAS17 anti-Ha-*ras* oligonucleotide targeted to the codon 12 mutation.

Tumour volume measurements were conducted periodically during the 35 days following cell injection. The results indicated strong inhibition of EJ tumour growth by RAS17, but also some non-specific inhibition by the scrambled sequence (Fig. 2).

Comparable antitumour results have been reported by two other groups (Schwab et al 1994, Bennett et al 1996). The efficacy and potency of RAS17 in cells and in mice provides a baseline for comparison between the two systems. DNA phosphorothioates complexed with Lipofectamine are approximately as potent in cell culture as free DNA phosphorothioates administered to mice. It is clear that the scrambled control also inhibits tumour growth to some extent, presumably by a non-antisense mechanism, such as protein binding. For example, phosphorothioates have been shown to bind to fibroblast growth factor 2, inhibiting its role in tumour promotion (Benimetskaya et al 1995).

Ki-ras-induced pancreatic cancer

Pancreatic cancer progresses rapidly and is usually refractory to treatment, with a median survival time of 12 months. For all stages, only 3% of patients live more than five years after diagnosis. Pancreatic cancer provides the most genetically clear-cut indication for antisense therapy; 90% of the incidence arises from a mutation in the 12th codon of Ki-*ras* (Almoguera et al 1988). The mutational spectrums of Ki-*ras* detected in colon, lung and pancreatic tumours have been characterized. For example, the G→T and G→A mutations in the second base of codon 12 account for 65% of the Ki-*ras* mutations observed in pancreatic tumours. Mutant Ki-*ras* may be detected in circulating cells, bile and pancreatic juice, as well as in biopsied tissue (Tada et al 1993). Therefore, genetic analysis of physiological samples would allow a rapid determination of the activating mutation in Ki-*ras*, which would facilitate the most appropriate choice of oligonucleotide therapy.

Complementary DNA phosphorothioates have been applied against Ki-*ras* (Ehlert et al 1997a), and the antisense sequence KRASSA (5′ dAGT CGC CCC GCC GCA 3′) was observed to decrease Ki-*ras* p21 levels in AsPC-1 human metastatic pancreatic adenocarcinoma cells treated with oligonucleotide–Lipofectamine complexes as above in a sequence- and dose-dependent manner (Fig. 3). Similarly, KRASSA displayed antiproliferative activity (Fig. 4) and inhibited tumorigenesis by Ki-*ras*-transformed AsPC-1 cells in nude mice (Fig. 5) in a sequence-specific and dose-dependent fashion. Under the conditions and oligonucleotide concentrations of these experiments, the sense control KRASSS did not display non-specific inhibition.

Comparable results were reported by Bennett et al (1996) with an unpublished sequence. Similarly, pancreatic cancer cell lines permanently transformed with an antisense construct displayed limited tumorigenicity (Aoki et al 1995). Hence, Ki-*ras* appears to be a logical target for oligonucleotide therapy of pancreatic cancer.

In the cell culture experiments approximately fivefold higher doses of oligonucleotides were required to obtain substantial antigen reduction than in the

| 1 | 2 | 3 | 4 | 5 | 6 | 7 | 8 | 9 | 10 |

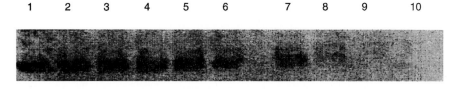

FIG. 3.   Inhibition of AsPC-1 human metastatic pancreatic adenocarcinoma cell Ki-*ras* p21 protein levels by DNA phosphorothioates. Following 16 h of exposure to oligonucleotide–Lipofectamine (1 : 6, w/w; Gibco/BRL) in Opti-Mem (Life Technologies, Gaithersburg, MD) at 37 °C, AsPC-1 cells were incubated for an additional 144 h in Dulbecco's modified Eagle's medium plus 10% fetal bovine serum. Cell lysates were processed for Western blot analysis. Lane 1, no treatment; lane 2, lipid only; lane 3, 0.5 µM KRASSS; lane 4, 1.0 µM KRASSS; lane 5, 1.5 µM KRASSS; lane 6, 2.0 µM KRASSS; lane 7, 0.5 µM KRASSA; lane 8, 1.0 µM KRASSA; lane 9, 1.5 µM KRASSA; lane 10, 2.0 µM KRASSA.

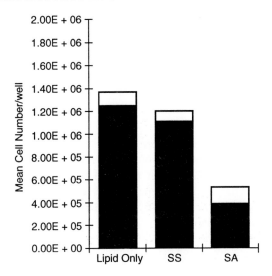

FIG. 4. Inhibition of AsPC-1 human metastatic pancreatic adenocarcinoma cell proliferation by Ki-*ras* phosphorothioate oligonucleotide–Lipofectamine (Gibco/BRL) complexes. Cells were treated as in Fig. 3, at 1.5 μM KRASSS (SS) or KRASSA (SA), and titred at the end of incubation. Cell number per well is shown on the *y*-axis. Unshaded area at the top of each bar

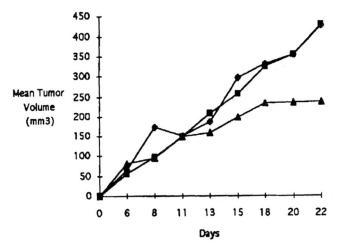

FIG. 5. Inhibition of tumour growth by subcutaneous DNA phosphorothioates. Tumours were induced in groups of five female BALB/c *nu/nu* mice by subcutaneous injection of 6 × 10⁵ AsPC-1 human metastatic pancreatic adenocarcinoma cells on Day 0. Subcutaneous DNA administration began on Day 1 on the flank opposite the site of AsPC-1 injection, with a dose of 167 nmol three times a week for 14 days, for a total of 1 μmol, or 5 mg. Tumour volumes were measured for each animal three times a week until sacrifice at Day 22. Symbols: ◆, PBS vehicle; ■, KRASSS sense control; ▲, KRASSA antisense oligomer.

case above of activated Ha-*ras* in bladder cancer cells. On the other hand, the cellular incubations were threefold longer, which may account for the difference. In the case of the Ki-*ras* oligonucleotides, no significant non-specific effects were observed from the control sequence.

erbB2-induced ovarian cancer

Invasive ovarian and breast cancer incidence rates for women continue to increase, despite recent advances in the treatment of advanced disease, screening for early cancer, and fundamental knowledge about the molecular and cellular events that underlie these diseases. The proto-oncogene c-*erbB2*, also called *neu* or HER-2, is amplified and/or overexpressed in about 30% of ovarian cancers (Slamon et al 1989, Berchuck et al 1990) and mammary tumours (Press et al 1993). Independent of gene copy number, overexpression of the *erbB2* p185 antigen positively correlates with poor prognosis (Berns et al 1995). The 185 kDa *erbB2* protein displays strong homology with epidermal growth factor receptor, part of the receptor tyrosine kinase family of cell surface proteins (Goldman et al 1990). Because the p185 antigen is overexpressed early and throughout the course of the disease, one may hypothesize that ablation of *erbB2* expression may specifically inhibit or regress ovarian and breast cancer. Indeed, a bispecific murine monoclonal antibody against ErbB2, 2B1, was observed to inhibit ascites, pleural effusion, chest wall disease and liver metastasis in four patients out of 15 in a Phase I clinical trial, although the murine antibody induced strong antigenic reactions, and briefly decreased platelet and leukocyte counts (Weiner et al 1995). The ErbB2 receptor does not appear to be expressed in normal adult cells, except for secretory epithelial cells, and it may only be necessary at an early developmental stage (Kokai et al 1987). Hence, ablation of *erbB2* expression is not predicted to be deleterious to most normal cells.

Complementary oligonucleotides targeted against *erbB2* message have been applied previously to breast cell lines, resulting in down-regulation of p185, the *erbB2* transcript and inhibition of cell proliferation (Bertram et al 1994, Colomer et al 1994, Vaughn et al 1995), leading to the identification of US3 (5′ dGGTGCTCACTGCGGC 3′) as an efficacious antisense sequence (Vaughn et al 1996). Because of the similar pathologies of mammary and ovarian adenocarcinomas, as well as the shared prognostic indicator of overexpressed p185 (Tyson et al 1991), we examined oligonucleotide efficacy in a tumorigenic subclone (ip1) of SKOV3 human ovarian adenocarcinoma cells (kind gift of M. C. Hung; Zhang et al 1995). Western blots revealed sequence-dependent reduction of p185 in SKOV3 cells treated with the anti-*erbB2* sequence US3 (Fig. 6), whereas counts of viable treated cells revealed inhibition of cell proliferation (Fig. 7; Ehlert et al 1997b). However, antiproliferation was equally induced by the scrambled control, suggesting that antiproliferation was not necessarily the result of sequence-dependent reduction of ErbB2 antigen.

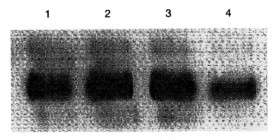

FIG. 6.   Inhibition of SKOV3 human ovarian adenocarcinoma cell *erbB2* p185 protein levels by DNA phosphorothioates. Following 16 h of exposure to oligonucleotide–Lipofectamine (1 : 6, w/w; Gibco/BRL) in Opti-Mem (Life Technologies, Gaithersburg, MD) at 37 °C, SKOV3 cells were incubated for an additional 144 h in Dulbecco's modified Eagle's medium plus 10% fetal bovine serum. Cell lysates were processed for Western blot analysis. Lane 1, no treatment; lane 2, lipid only; lane 3, 1.0 μM US3 (5′ dGGTGCTCACTGCGGC 3′); lane 4, 1.0 μM SC3 (scrambled control).

The tumorigenicity of the SKOV3.ip1 xenografts was confirmed here, reproducibly displaying significant tumour volumes within 28 days to allow assessment of oligonucleotide inhibition of tumorigenesis. Nude mice implanted with SKOV3.ip1 cells were treated by continuous slow release from subcutaneously implanted Alzet micropumps (Alza, Palo Alto, CA) (Fig. 8). Inhibition of tumour growth by US3

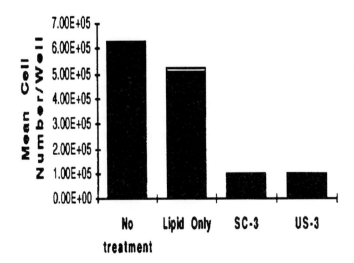

FIG. 7.   Inhibition of SKOV3 human ovarian adenocarcinoma cell proliferation by *erbB2* phosphorothioate oligonucleotide–Lipofectamine (Gibco/BRL) complexes. Cells were treated as in Fig. 6, and titred at the end of the incubation. Cell number per well is shown on the *y*-axis. Unshaded area at the top of lipid-only bar represents S.E.M. SC-3, scrambled control; US-3, 5′ dGGTGCTCACTGCGGC 3′.

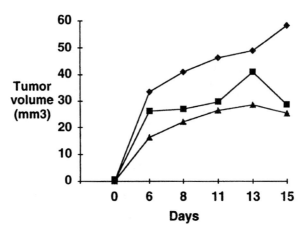

FIG. 8.   Inhibition of tumour growth by subcutaneous DNA phosphorothioates. Tumours were induced in groups of 12–15 female BALB/c *nu/nu* mice by subcutaneous injection of 1×10^6 SKOV3 human ovarian adenocarcinoma cells on Day 0. Alzet 2002 micropumps (Alza, Palo Alto, CA) were implanted subcutaneously on Day 1 on the flank opposite the site of SKOV3 injection, continuously releasing 2.5 nmol/h, or 60 nmol/day, for 14 days, for a total of 1 μmol, or 5 μg. Tumour volumes were measured for each animal three times a week until sacrifice at Day 15. Symbols: ◆, phosphate-buffered saline vehicle; ■, SC3 (scrambled control); ▲, US3 antisense oligomer 5′ dGGTGCTCACTGCGGC 3′.

was clearly demonstrated, but SC3 was equally effective. Thus, as in the antiproliferation experiment (Fig. 7), tumour inhibition was not sequence specific.

The regimen and dose rate used in Fig. 8 leads to a steady-state serum oligonucleotide concentration, measured by OliGreen™, of 0.1 μM (Gray & Wickstrom 1997). Thus, antitumour efficacy was observed in mice at a serum concentration 10-fold less than that which effectively inhibited cell proliferation in culture. Given the tumour distribution of phosphorothioates (DeLong et al 1997), oligonucleotide concentration in the tumour tissue was much less than 0.1 μM. With the continuous release mode of administration, cationic lipid complexation was not necessary.

c-myc-induced lymphoma

Activation of c-*myc*, the expression of which is required for cell proliferation, has been implicated in a number of leukaemias and solid tumours in a variety of mammalian species (Bishop 1991). For example, Burkitt's lymphoma is a frequent haematological malignancy of childhood. This malignancy, like all others, results from a multistep process of oncogene activation or suppressor gene inactivation. Virtually all cases of Burkitt's lymphoma display translocation of chromosomes 8

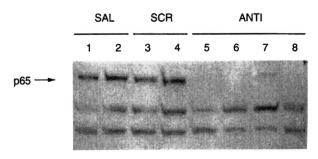

FIG. 9. Western blot analysis of Myc antigen expression in spleens of individual Eµ-*myc* transgenic mice with palpable tumours, treated with saline (SAL) (lanes 1, 2), scrambled (SCR) (lanes 3, 4) or MYC6 antisense (ANTI) (lanes 5–8) DNA phosphorothioates for one week. Each well contained 10 µg of nuclear protein extract. Each lane represents a different mouse.

and 14, placing c-*myc* from chromosome 8 under the transcriptional control of the immunoglobulin M heavy chain enhancer from chromosome 14 (Bishop 1991).

c-*myc* is a frequent target for complementary DNA inhibition and resultant inhibition of cellular proliferation (Heikkila et al 1987). *In vitro* studies using human complementary c-*myc* oligonucleotides targeted to various sites in c-*myc* mRNA have been reported to inhibit proliferation of leukaemic cells, breast cancer cells and smooth muscle cells (Huang et al 1995). The antiproliferative effects correlated with a 50–93% reduction in the level of Myc, at oligonucleotide culture medium concentrations of 5–10 µM (Wickstrom et al 1988). Translating this approach to a murine model bearing a murine immunoglobin enhancer/c-*myc* fusion transgene (Eµ-*myc*), a single administration of complementary c-*myc* DNA was observed to down-regulate c-*myc* expression in splenic and peripheral lymphocytes (Wickstrom et al 1992).

To determine the effects of continuous oligonucleotide therapy over an extended period (Huang et al 1995), we treated Eµ-*myc* mice prophylactically for six weeks, beginning at the time of weaning, with a complementary c-*myc* phosphorothioate oligonucleotide, MYC6 (5' dCACGTTGAGGGGCAT 3'), or a scrambled control. The oligonucleotides, or PBS vehicles, were released continuously at 2.5 nmol/h from a subcutaneously implanted Alza micropump, resulting in a steady-state serum concentration of 0.1 µM (Gray & Wickstrom 1997). This regimen ablated Myc antigen levels in the spleens of Eµ-*myc* mice treated with MYC6 (Fig. 9).

Among the MYC6 cohort, 75% were still free of tumours at the age of 26 weeks, whereas half of the mice treated with a scrambled control DNA or saline displayed palpable tumours by eight to nine weeks and 95% developed tumours by 16 weeks (Fig. 10). Subsequent studies in a residual disease transplant model have shown comparable or greater potency by other anti-c-*myc* sequences in Myc antigen reduction and tumour inhibition (Smith & Wickstrom 1997).

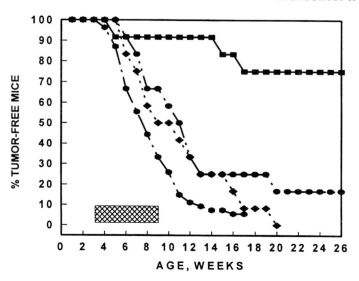

FIG. 10.   Age of onset of lymphoid tumours in Eμ-*myc* transgenic mice treated with saline, scrambled or MYC6 antisense DNA phosphorothioates. Hatched box shows period of continuous administration of DNA from micro-osmotic pumps. ●, no treatment (54 mice which were tumour free at three weeks); ◆, saline (12 mice); ●, scrambled DNA phosphorothioate (12 mice); ■, MYC6 antisense DNA phosphorothioate (12 mice).

Again, one sees 100-fold to 1000-fold potency in the mouse model, compared to the cell culture system. This situation is most beneficial for the application of oligonucleotide therapy in the clinic.

Summary

Contrary to expectation, a number of therapeutic oligonucleotides have displayed greater potency in animals than in cell culture. The data do not support the hypothesis of random toxicity by all oligonucleotide sequences contributing to antitumour effects, although this may in fact be the case for the *erbB2* system studied above. The natural carrier hypothesis, however, has not been examined, nor has the possibility that cells in living tissue take up oligonucleotides much more avidly than cells in culture.

How could one identify the postulated naturally occurring vehicles for oligonucleotide distribution and cellular uptake which parallel the role of cationic lipids often used in cell culture? First of all, size exclusion may be examined. Small samples of cells may be cultured in sealed permeable hollow fibres implanted subcutaneously in mice. If oligonucleotide potency in this system parallels that observed in tumour inhibition experiments, then the putative murine factors which elevate oligonucleotide potency are small enough to penetrate the hollow fibres. A negative result eliminates the latter hypothesis, and allows the possible involvement of soluble factors, carrier macromolecules, too large to permeate the hollow fibres. One may also hypothesize that malignant cells growing on a tissue substrate remodel their surfaces, relative to the cell culture situation, to promote uptake of extracellular compounds, including oligonucleotides. It may be possible to test this model by administering small doses of fluorescent oligonucleotides, both to cells in culture and to mice carrying implanted cells, and then examining cellular uptake and trafficking of fluorescent oligonucleotides over time.

The examples described above were limited to DNA phosphorothioates. Perhaps similar experiences will be found with some of the other DNA derivatives available for study. In any case, the phenomenon of greater oligonucleotide potency in animals than in culture dishes represents one more fortunate development in the history of oligonucleotide therapy.

Acknowledgements

We are indebted to our colleagues who participated in these investigations: Grace Ehlert, Gary Gray, Irina Feldman, Hitoshi Hayasaka, Ye Huang, Michael Kligshteyn, Susan Sladon and Janet Smith. We are also grateful to our colleagues who provided critical analysis and useful suggestions: Marshall Anderson, Marvin Caruthers, Dirk Iglehart and Rudolph Juliano. This work was supported by American Cancer Society grant DHP-105, and National Institutes of Health grants CA42960 and CA60139, to E. W.

References

Agrawal S 1996a Antisense therapeutics. Humana Press, Totowa, NJ

Agrawal S 1996b Antisense oligonucleotides: toward clinical trials. Trends Biotechnol 14:376–387

Akhtar S 1995 Delivery strategies for antisense oligonucleotide therapeutics. CRC Press, Boca Raton, FL

Almoguera C, Shibata D, Forrester K, Martin J, Arnheim P, Perucho M 1988 Most human carcinomas of the exocrine pancreas contain mutant c-K-Ras genes. Cell 53:549–554

Aoki K, Yoshida T, Sugimura T, Terada M 1995 Liposome-mediated *in vivo* gene transfer of antisense K-*ras* construct inhibits pancreatic tumor dissemination in the murine peritoneal cavity. Cancer Res 55:3810–3816

Belikova AM, Zarytova VF, Grineva NI 1967 Synthesis of ribonucleosides and diribonucleoside phosphates containing 2-chloroethylamine and nitrogen mustard residues. Tetrahedron Lett 37:3557–3562

Benimetskaya L, Tonkinson JL, Koziolkiewicz M et al 1995 Binding of phosphorothioate oligodeoxynucleotides to basic fibroblast growth factor, recombinant soluble CD4, laminin and fibronectin is P-chirality independent. Nucleic Acids Res 23:4239–4245

Bennett CF, Dean N, Ecker DJ, Monia BP 1996 Pharmacology of antisense therapeutic agents. In: Agrawal S (ed) Antisense therapeutics. Humana Press, Totowa, NJ, p 13–46

Berchuck A, Kamel A, Whitaker R et al 1990 Overexpression of HER-2/neu is associated with poor survival in advanced epithelial ovarian cancer. Cancer Res 50:4087–4091

Berns EMJJ, Foekens JA, van Staveren IL et al 1995 Oncogene amplification and prognosis in breast cancer: relationship with systemic treatment. Gene 159:11–18

Bertram J, Killian M, Brysch W, Schlingensiepen K-H, Kneba M 1994 Reduction of *erbB2* gene product in mammary carcinoma cell lines by *erbB2* mRNA-specific and tyrosine kinase consensus phosphorothioate antisense oligonucleotides. Biochem Biophys Res Commun 200:661–667

Bishop JM 1991 Molecular themes in oncogenesis. Cell 64:235–248

Colomer R, Lupu R, Bacus SS, Gelmann EP 1994 *erbB2* antisense oligonucleotides inhibit the proliferation of breast cancer carcinoma cells with ErbB2 oncogene amplification. Br J Cancer 70:819–825

Crooke ST, Lebleu B 1993 Antisense research and applications. CRC Press, Boca Raton, FL

Daaka Y Wickstrom E 1990 Target dependence of antisense oligodeoxynucleotide inhibition of c-Ha-*ras* p21 expression and focus formation in T24-transformed NIH3T3 cells. Oncogene Res 5:267–275

DeLong RK, Nolting A Fisher M et al 1997 Comparative pharmacokinetics, tissue distribution and tumor accumulation of phosphorothioate, phosphorodithioate, and methylphosphonate oligonucleotides in nude mice. Antisense Nucleic Acid Drug Dev 7:71–77

Ehlert MG, Sladon SE, Wickstrom E, Anderson MW, Tyson FL 1997a Response of human pancreatic tumor cells to K-*ras* phosphorothioate antisense oligonucleotides, submitted

Ehlert MG, Sladon SE, Wickstrom E, Tyson FL 1997b Proliferative responses of human ovarian adenocarcinoma cells to HER-2/neu phosphorothioate antisense oligonucleotides, submitted

Goldman R, Ben-Levy R, Peles E, Yarden Y 1990 Heterodimerization of the ErbB-1 and ErbB2 receptors in human breast carcinoma cells: a mechanism for receptor transregulation. Biochemistry 29:11024–11028

Gray GD, Wickstrom E 1997 Rapid measurement of modified oligonucleotides in plasma using a single-stranded DNA binding fluorophore. Antisense Nucleic Acid Drug Dev 7:133–140

Gray GD, Hebel D, Hernandez O, Root M, Pow-Sang JM, Wickstrom E 1993 Antisense DNA inhibition of c-Ha-*ras* induced tumor growth in nude mice. Cancer Res 53:577–580

Heikkila R, Schwab G, Wickstrom E et al 1987 A c-*myc* antisense oligodeoxynucleotide inhibits entry into S phase but not progress from G0 to G1. Nature 328:445–449

Huang Y, Snyder R, Kligshteyn M, Wickstrom E 1995 Prevention of tumor formation in a mouse model of Burkitt's lymphoma by six weeks of treatment with anti-c-*myc* DNA phosphorothioate. Mol Med 1:647–658

Kokai Y, Cohen JA, Drebin JA, Greene MI 1987 Stage and tissue-specific expression of the neu oncogene in rat development. Proc Natl Acad Sci USA 84:8498–8501

Lowy DR, Willumsen BM 1993 Function and regulation of Ras. Annu Rev Biochem 62:851–891

Press MF, Pike MC, Chazin VR et al 1993 Her-2/neu expression in node-negative breast cancer: direct tissue quantitation by computerized image analysis and association of overexpression with increased risk of recurrent disease. Cancer Res 53:4960–4970

Schwab G, Chavany C, Durooux I et al 1994 Antisense oligonucleotides adsorbed to polyalkylcyanoacrylate nanoparticles specifically inhibit mutated Ha-*ras*-mediated cell proliferation and tumorigenicity in nude mice. Proc Natl Acad Sci USA 91:10460–10464

Slamon DJ, Godolphin W, Jones LA et al 1989 Studies of Her-2/neu proto-oncogene in human breast and ovarian cancer. Science 244:707–712

Smith JB, Wickstrom E 1997 Inhibition of tumorigenesis in a mouse model of B-cell lymphoma by c-*myc* complementary oligonucleotides. Proc Am Assoc Cancer Res 38:316

Tabin CJ, Bradley SM, Bargmann CI et al 1982 Mechanism of activation of a human oncogene. Nature 300:143–149

Tada M, Omata M, Shigenobu K, Asisho H, Ohto M, Saiki K, Sninsky JJ 1993 Detection of Ras gene mutations in pancreatic juice and peripheral blood of patients with pancreatic adenocarcinoma. Cancer Res 53:2472–2474

Tyson FL, Boyer CM, Kaufman R et al 1991 Expression and amplification of the HER-2/neu (c-*erbB-2*) protooncogene in epithelial ovarian tumors and cell lines. Am J Obstet Gynecol 165:640–646

Vaughn JP, Iglehart JD, Demirdji S et al 1995 Antisense DNA downregulation of the ERBB2 oncogene measured by a flow cytometric assay. Proc Natl Acad Sci USA 92:8338–8342

Vaughn JP, Stekler J, Demirdji S et al 1996 Inhibition of the *erbB-2* tyrosine kinase receptor in breast cancer cells by phosphoromonothioate and phosphorodithioate antisense oligonucleotides. Nucleic Acids Res 24:4558–4564

Weiner LM, Clark JI, Davey M et al 1995 Phase I trial of 2B1, a bispecific monoclonal antibody targeting c-*erbB2* and FcgRIII. Cancer Res 55:4586–4593

Wickstrom E 1991 Prospects for antisense nucleic acid therapy of cancer and AIDS. Wiley, New York

Wickstrom E 1997 Clinical trials of genetic therapy with antisense DNA and DNA vectors. Marcel Dekker, New York, in press

Wickstrom EL, Bacon TA, Gonzalez A, Freeman DL, Lyman GH, Wickstrom E 1988 Human promyelocytic leukemia HL-60 cell proliferation and c-*myc* protein expression are inhibited by an antisense pentadecadeoxynucleotide targeted against c-*myc* mRNA. Proc Natl Acad Sci USA 85:1028–1032

Wickstrom E, Bacon TA, Wickstrom EL 1992 Down-regulation of c-*myc* antigen expression in lymphocytes of Eμ-c-*myc* transgenic mice treated with anti-c-*myc* DNA methylphosphonate. Cancer Res 52:6741–6745

Zamecnik PC, Stephenson ML 1978 Inhibition of Rous sarcoma virus replication and cell transformation by a specific oligodeoxynucleotide. Proc Natl Acad Sci USA 75:280–284

Zhang Y, Yu D, Xia W, Hung M-C 1995 HER-2/neu-targeting cancer therapy via adenovirus-mediated E1A delivery in an animal model. Oncogene 10:1947–1953

DISCUSSION

Stein: Your data on tetraplex formation at 37 °C seem to be at variance with those of Karen Fearon. Can you clarify this?

Fearon: At room temperature, when we dissolved LR-3280 (MYC6) in saline we observed by size-exclusion chromatography an approximate 60 : 40 peak-area ratio of monomer to tetraplex, which presumably corresponds to an approximately 85 : 15 molar ratio, due to the approximately fourfold higher molar extinction coefficient of the tetraplex. We've also looked at 90 °C and found that it disaggregates within 2 min, but I don't know what the behaviour is at 37 °C. One of the experiments we want to do is see if the tetraplex can act as a reservoir for the monomeric form in the presence of the target RNA.

Stein: Your gels were interesting in the sense that the second band could be a duplex, and the duplex could be in equilibrium with the tetraplex.

Wickstrom: It would have to be a homoduplex formed via G–G interactions. It's possible that the G_4 sequences bind to some cell surface or cytosolic protein, and that the G_4 sequences can act either as a monomer or as a homoduplex. The only conclusion we can draw is that MYC6 (LR 3280) is probably not acting as a tetraplex.

Stein: What happens when you boil your samples?

Wickstrom: Five minutes at 90 °C results in the appearance of the monomeric form only. This was the starting point for all the experiments before the incubations at 4 °C, 23 °C and 37 °C.

Stein: This demonstrates that the flanking sequences have an effect because we have some oligonucleotides that have the same motif, but if we boil them we don't observe the monomeric form, unless we also treat them with formamide and put them on 10 M urea sequencing gels. If we put them on 7 M urea sequencing gels we observe tetraplexes.

Wickstrom: Sudhir Agrawal showed that the location of the G_4 sequence has a large impact on the level of hyperstructure formation (Agrawal et al 1996). The MYC6 sequence 5' dAACGTTGAGGGGCAT 3' seems to escape significant tetraplex formation, and the MYC5 sequence 5' dCTCGTCGTTTCCTCA 3' is even more effective, but it doesn't have a G_4 sequence.

Cohen: There is also the possibility that even if a G_4 structure is not observed intrinsically it can be induced upon binding to a particular receptor or site. On the other hand, whether or not a G_4 sequence is present in an oligomer may be irrelevant to its biological effect because its structure may fall apart in the cell.

Gewirtz: Has anybody looked at tetraplex uptake versus the monomeric form?

Agrawal: In certain cell lines the tetraplex is taken up more rapidly.

Gewirtz: Is it possible, therefore, that if the tetraplex is destroyed, the uptake is reduced and the effect is lost?

Stein: It is possible, but it is probably oligonucleotide restricted and cell type restricted, at least to some extent. We've looked at the uptake of model tetraplexes, so they're not the same sequences that you have been working on. However, we have had problems with this because if the G_4 motif is at the 5' end and we put a tag on the 5' terminus then the rate at which the tetraplex forms changes.

Crooke: We didn't see significant cell uptake of ISIS 5320 dTTGGGGTT, which is an 8 mer with a G_4 sequence, in the cell lines that we looked at, although we did see cell adsorption. It's difficult to discriminate between these processes because the phosphorothioate G_4 oligonucleotides have a high avidity for many types of protein. It's also difficult to generalize across different G_4 sequences in different cells. This problem is exaggerated in animals.

Akhtar: We have a sequence that forms higher structures, and its cellular uptake is enhanced to a level that is normally expected for cholesterol-modified oligonucleotides. Hence, it is likely that guanosine-rich sequences exhibit altered pharmacokinetic/pharmacodynamic properties.

Agrawal: We found that the pharmacokinetics of G_4 oligonucleotides *in vivo* have an impact on the plasma clearance and tissue distribution, which are significantly different in oligonucleotides without G_4.

Crooke: When we did the ISIS 5320 studies in the SCID mouse model, we spent a lot of time demonstrating that after we administered it a large fraction of it was still tetraplex. We observed different pharmacokinetic properties from the single-stranded phosphorothioate, and we also observed activity in that SCID mouse model.

Cohen: In your *in vitro* studies have you considered that phosphorothioates bind to plastic surfaces (Watson et al 1992), so that you may be losing a lot of material?

Wickstrom: The medium is placed in the dish before the experiment, so I would imagine that the dish is already coated with protein. One interesting test that's going to be performed by Melinda Hollingshead at the National Cancer Institute Frederick Cancer Research Facility, Maryland, is to grow some of these lines in hollow fibres inside mouse hosts. They will be looking at a sort of hybrid system where the cells are not sitting on a tissue matrix bathed in cellular fluids, rather in a hollow fibre which has some degree of size exclusion so that the effects of diffusible agents can be studied, versus the effects of the tissue matrix.

Nicklin: Is your time-averaged tumour concentration also a distribution-averaged tumour concentration?

Wickstrom: Yes. In a collaboration with Rudy Juliano, University of North Carolina, we have looked at the distribution of phosphorothioates, methylphosphonates and dithioates in tumours, and the tumour had the same distribution fraction as muscle (DeLong et al 1997).

Nicklin: But if you did autoradiography across the tumour, you would see moderate levels of oligonucleotide throughout the tumour, whereas in proliferating areas there would be a high concentration of oligonucleotide, so it is misleading to average the two concentrations.

Wickstrom: That's a valuable observation because we often see a lack of total tumour ablation in various systems, which suggests that we have to improve the permeation of the tumour by the oligonucleotides.

Monia: You mentioned using a Ha-*ras* antisense oligonucleotide for Ha-*ras*-transformed tumours and a K-*ras* antisense oligonucleotide for K-*ras*-transformed tumours. We've observed that for a pancreatic tumour line which contains a K-*ras* mutation Ha-*ras* antisense oligonucleotides are far more effective than K-*ras* antisense oligonucleotides. Have you looked at the effect of your Ha-*ras* oligonucleotide on the pancreatic tumour that you studied?

Wickstrom: No. That approach involves looking at targets other than the original activating oncogene to demonstrate broad spectrum activity, and we haven't done any of those sorts of experiments.

Iversen: What is the rationale for thinking that *erbB2* can be used as antitumour agent? *erbB2* is overexpressed, but is it actually involved in the cell viability of those tumours and is there an antisense effect?

Wickstrom: The antisense experiments that we performed were designed to test that possibility, and it appears that the receptor is not the key to the survival of those cells. It is a marker that correlates with poor prognosis, so it was reasonable to think that an overexpressed receptor would play a role in proliferation, but the cell culture proliferation and tumour growth experiments suggest that this is not the case.

Hélène: When HL-60 cells are treated with an antisense oligonucleotide targeted to c-*myc*, for example, the HL-60 cells differentiate (Wickstrom et al 1989). When T24 cells are treated with *ras* antisense oligonucleotides the cells stop proliferating. If decreases in c-*myc* or *ras* mRNAs are observed, how can one be sure that this is due to the direct effect of the oligonucleotide, rather than the arrest of cell proliferation or the induction of differentiation?

Monia: c-*myc* is particularly interesting because it is regulated by proliferation, and we have also observed that *ras* mRNA levels are down-regulated to some extent when cell growth is stationary. Therefore, this is a possibility, and the best controls are those that incorporate mismatches such that a graded effect on mRNA expression is observed.

Crooke: Another approach is to do time course experiments. We looked at acute reductions in RNA levels occurring before changes in cell proliferation either *in vitro* or *in vivo,* and we found that decreases in RNA levels occurred prior to cell population changes only in some cases. For a long time we have wrestled with the meaning of those reductions in the mRNA levels.

Gewirtz: It's also not strictly correct that when proliferation is halted the levels of c-*myc* mRNA decrease immediately. There are many experimental systems in which c-*myc* expression stays elevated for hours or days. As long as c-*myc* expression is elevated the cells can be rescued with serum factors, for instance, and they will continue to divide. When the levels of c-*myc* finally do decrease the cells cannot be rescued.

In the transgenic mouse models you were using the material prophylactically to prevent the growth of a spontaneous tumour. Did you also show that c-*myc* RNA and protein levels were decreasing in these animals?

Wickstrom: The levels of mRNA don't decrease as much as the protein levels.

Gewirtz: If the animals are developing with low levels of c-*myc* protein how do they manage to make an immune system or blood? Because non-tumour cells will still need to express c-*myc* protein.

Wickstrom: Normal non-transgenic mice don't need much c-*myc* protein to grow normal tissues.

Gewirtz: But presumably those low levels of c-*myc* protein are being down-regulated as well.

Wickstrom: That's not necessarily the case. Transformed cells are typically much more aggressive at taking up oligonucleotide (Gray et al 1997), and even many years ago we observed that normal peripheral blood lymphocytes were much less sensitive to c-*myc* oligonucleotide than transformed HL-60 cells (Wickstrom et al 1988).

Gewirtz: But transgenic mice are different because every cell expresses c-*myc*.

Wickstrom: Not necessarily, because the immunoglobulin heavy chain enhancer is activated in B cells.

Gewirtz: But what about the lymphoid cells that normally become activated during the animal's development?

Krieg: The enhancer should be turned on in any activated B cells, so a prediction is that responses to a B cell mitogen, such as lipopolysaccharide, in terms of proliferation or isotype switching would likely be altered in the transgenic mice. One might also predict true that T cell development, for example, is normal in those mice to the extent that the enhancer is cell type specific.

Gewirtz: It would be interesting to challenge those mice and see what they do.

Krieg: Or try to vaccinate them and see if they can make an antigen-specific response.

Crooke: I would like to address the surprising potency in animals, compared to the situation *in vitro* which does not relate strictly to tumours. In a wide range of models of all sorts of diseases with a variety of different oligonucleotides targeted to different mRNAs we have seen surprising potencies. Therefore, it's a broad issue. I am reminded of the study of aspirin, of which there are tens of thousands of papers, and I don't believe there's a proven mechanism of uptake of this compound into cells. I am confident that there are no data on sub-organ distribution, and there is no adequate explanation for the extraordinary potency of aspirin *in vivo* and the duration of its inhibition of prostaglandin synthases. Finally, there's no explanation for why the gastric mucosa of dogs is 50-fold more sensitive to aspirin than human gastric mucosa. Therefore, we shouldn't be too discouraged if in two years from now we haven't solved the oligonucleotide potency problem.

References

Agrawal S, Iadarola PL, Temsamani J, Zhao QY, Shaw DR 1996 Effect of G-rich sequences on the synthesis, purification, hybridization, cell uptake, and hemolytic activity of oligonucleotides. Bioorg Med Chem Lett 6:2219–2224

DeLong RK, Nolting A, Fisher M et al 1997 Comparative pharmacokinetics, tissue distribution and tumor accumulation of phosphorothioate, phosphorodithioate, and methylphosphonate oligonucleotides in nude mice. Antisense Nucleic Acid Drug Dev 7:71–77

Gray GD, Basu S, Wickstrom E 1997 Transformed and immortalized cellular uptake of oligodeoxynucleoside phosphorothioates, 3' alkylamino oligodeoxynucleotides, 2'-O-methyl oligoribonucleotides, oligodeoxynucleoside methylphosphonates, and peptide nucleic acids. Biochem Pharmacol 53, in press

Watson PH, Pon RT, Shiu RP 1992 Inhibition of cell adhesion to plastic substratum by phosphorothioate oligonucleotide. Exp Cell Res 202:391–397

Wickstrom EL, Bacon TA, Gonzalez A, Freeman DL, Lyman GH, Wickstrom E 1988 Human promyelocytic leukemia HL-60 cell proliferation and c-*myc* protein expression are inhibited by an antisense pentadecadeoxynucleotide targeted against c-*myc* mRNA. Proc Natl Acad Sci USA 85:1028–1032

Wickstrom EL, Bacon TA, Gonzalez A, Lyman GH, Wickstrom E 1989 Anti-c-*myc* DNA oligomers increase differentiation and decrease colony formation by HL-60 cells. In Vitro Cell Dev Biol 26:297–302

Structure–activity relationships in cell culture

Richard W. Wagner

Gilead Sciences Inc., 353 Lakeside Drive, Foster City, CA 94404, USA

Abstract. The use of antisense oligonucleotides in cell culture relies on the development of potent modifications and cell delivery techniques. C-5 propyne pyrimidine phosphorothioate oligonucleotides bind selectively and with high affinity to RNA within cells, leading to potent antisense inhibition. The effect that increasing steric bulk of C-5-substituted deoxyuridine analogues has on the affinity for RNA and the ability to inhibit gene expression is discussed. The GS 2888 cytofectin agent delivers oligonucleotides to cells at high efficiency in the presence of serum in cell media. Modifications leading to the discovery of GS 2888 centred on the aliphatic chain length of the molecule and the pK_a of the polar head group. Together, the C-5 propyne modifications and the GS 2888 cytofectin agent have been shown to be effective inhibitors of gene expression in cell culture, particularly in the area of cell cycle proteins involved in cancer progression.

1997 Oligonucleotides as therapeutic agents. Wiley, Chichester (Ciba Foundation Symposium 209) p 142–157

Antisense gene inhibition occurs when an oligonucleotide has sufficient access to its target RNA, binding affinity, stability to nucleases and permeability to cellular membranes such that it enters a cell and prevents translation (reviewed in Milligan et al 1993, Stein & Cheng 1993, Matteucci & Wagner 1996). Recently it was shown that potent gene-specific antisense inhibition was achieved in mammalian cells using C-5 propynyl pyrimidine (C-5 propyne) 2' deoxyphosphorothioate oligonucleotides (Fig. 1) (Wagner et al 1993). These oligonucleotides had the required binding affinity and nuclease stability properties to yield potent antisense inhibition when microinjected into cells. Inhibition was due to an RNase H-mediated mechanism.

The enhancement in binding affinity for complementary RNA by the C-5 propyne group is likely due to increased base stacking interactions between neighbouring base pairs. Enhanced affinity is a result of a substantially decreased dissociation rate of oligonucleotide/RNA duplexes Also, the C-5 propyne group does not interfere with RNase H cleavage kinetics. In gene inhibition studies these two properties — enhanced stability and RNase H cleavage — allow for the functional inactivation of RNA when oligonucleotide/RNA duplexes are microinjected into cells. This result

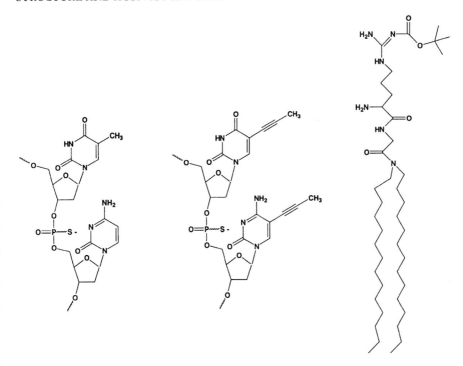

FIG. 1. Chemical structures of a phosphorothioate dimer containing unmodified T and C (left), and 5-(1-propynyl)uridine and 5-(1propynyl)cytosine (middle), and the GS 2888 lipid (right).

contrasts injected phosphorothioate/RNA duplexes that do not contain the C-5 propyne group: despite the observation that these duplexes form relatively stable structures, as measured *in vitro,* complete gene inhibition has not been observed when oligonucleotide/RNA duplexes are injected into cells. We previously proposed that an oligonucleotide/RNA unwinding activity may aid dissociation of non-propyne oligonucleotide/RNA duplexes in cells (Moulds et al 1995).

C-5 propyne phosphorothioate oligonucleotides do not readily permeate cells in culture (Wagner et al 1993). Oligonucleotides that show potent activity in microinjection experiments show no activity when incubated with cells in culture medium at high concentration (up to 100 μM for 48 h). Cell delivery methods, such as microinjection, electroporation and cationic lipid formulation techniques, work well in cell culture for delivering the oligonucleotides to cells, such that highly controlled antisense inhibition is observed (Flanagan & Wagner 1997). As discussed further below, we identified a cationic lipid formulation that showed superior oligonucleotide delivery properties compared to commercially available agents. This agent, GS 2888 cytofectin, delivers high quantities of oligonucleotides to the nuclei of cells in the presence of serum in the cell medium (Lewis et al 1996). The use of serum in experiments reduces much of the toxicity observed with cationic lipid delivery.

The development of successful antisense technology depends on the ability to rationally design potent inhibitors, and to demonstrate that the biological and clinical benefits of the oligonucleotide can be attributed to an antisense mechanism of action. Below I describe our experience using the C-5 propyne antisense and GS 2888 cytofectin agents. The effect that increasing steric bulk of C-5-substituted deoxyuridine analogues has on affinity for RNA and ability to inhibit gene expression is discussed. Modifications leading to the discovery of GS 2888 centred on the aliphatic chain length of the molecule and the pK_a of the polar head group. Also, rules for finding active sequences have emerged. We have focused on two general areas, namely, cell-based reporter systems and endogenously expressed cell cycle genes. Data generated thus far show that C-5 propyne oligonucleotides are a potent, promising class of antisense agents.

Antisense assays using microinjection

A useful approach for evaluating antisense effects is a nuclear microinjection assay in which African green monkey CV-1 kidney cells are injected with plasmids expressing mutant SV40 TAg and *Escherichia coli* β-galactosidase with or without oligonucleotides targeted to one of the proteins (Hanvey et al 1992, Wagner et al 1993). Immunofluorescence can be utilized to detect specific antisense gene inhibition using the non-targeted protein as an internal control. IC_{50} values (inhibitory concentration showing a 50% reduction in protein expression) are estimated by scoring for targeted versus non-targeted protein levels in cells relative to control injections containing no oligonucleotide.

This assay was initially utilized to demonstrate the antisense effects of peptide nucleic acids (PNAs) targeted to SV40 TAg (Hanvey et al 1992, Bonham et al 1995). Subsequent studies have utilized the system for demonstrating the potency of C-5 propyne phosphorothioate oligonucleotides targeted to either TAg or β-galactosidase (Wagner et al 1993, Moulds et al 1995, Wagner et al 1996). The system has been useful for showing that oligonucleotides which can act by an RNase H mechanism are more effective than those which act by an RNase H-independent mechanism (Moulds et al 1995).

Structure–activity relationships surrounding the C-5 propyne moiety

C-5 alkyne and C-5 thiazole dU analogues have been prepared and incorporated into an oligonucleotide targeted to the coding region of TAg (Fig. 2) (Gutierrez et al 1997). The stabilities of the C-5-substituted dU phosphorothiate oligonucleotide/ RNA complexes were determined by thermal melting temperature analysis. The propyne-substituted oligonucleotide showed a comparable increase in T_m, relative

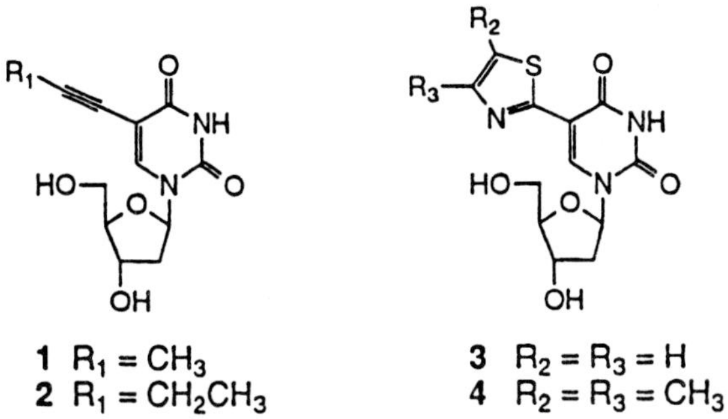

1 $R_1 = CH_3$
2 $R_1 = CH_2CH_3$

3 $R_2 = R_3 = H$
4 $R_2 = R_3 = CH_3$

FIG. 2.   C-5 2′-deoxyuridine analogues.

to the C-5 butynyl-substituted dU oligonucleotide, over the thymidine control ($\Delta T_m = +10.5\,^\circ C$ and $+10.0\,^\circ C$, respectively), and thiazole-substituted dU oligonucleotides and dimethylthiazole-substituted dU oligonucleotides formed the most stable duplexes ($\Delta T_m = +11.5\,^\circ C$ and $+13.0\,^\circ C$, respectively) (Gutierrez et al 1997).

Each oligonucleotide was analysed in the nuclear microinjection assay, and they all showed gene-specific inhibition of TAg relative to the β-galactosidase control. The C-5 propyne-substituted dU oligonucleotide had the most potent activity ($IC_{50} = 0.25\,\mu M$), followed by the C-5 butyne-substituted dU oligonucleotide ($1\,\mu M$), and the thiazole and dimethylthiazole-substituted dU oligonucleotides ($5\,\mu M$) (Table 1). The 20-fold decrease in activity for the thiazole-containing oligonucleotides was surprising due to a lack of correlation between activity and stability, as measured by T_m analysis, and warranted further investigation (Gutierrez et al 1997).

Association and dissociation kinetic analyses of the C-5-substituted dU oligonucleotides were evaluated in *in vitro* assays (Table 1). All of the oligonucleotides had comparable association kinetics, and, as indicated by the T_m analysis, the thiazole-containing oligonucleotides had the slowest dissociation rates. These data demonstrated that the thiazole-substituted oligonucleotides form the most thermodynamically stable duplexes with RNA (Gutierrez et al 1997).

Next, the RNase H cleavage kinetics for oligonucleotide/RNA duplexes were evaluated (Table 1). Each of the C-5-substituted dU oligonucleotides was bound to RNA and the relative rates of cleavage were performed using RNase H in HeLa nuclear extracts. C-5 propyne- and C-5 butyne-substituted oligonucleotides showed comparable cleavage efficiencies; however, cleavage of the thiazole-containing oligonucleotides was slower (approximately twofold for the thiazole dU oligonucleotide and fivefold for the dimethyl thiazole dU oligonucleotide). Thus, the decreased rate of RNase H cleavage may account, in part, for the decreased activity of the thiazole-substituted oligonucleotides (Gutierrez et al 1997).

A direct test of the ability of an oligonucleotide to inhibit gene expression is to preform the oligonucleotide onto a full length RNA and test whether the oligonucleotide inhibits translation of the RNA within the cell. Modified uridine-containing oligonucleotides were incubated with TAg RNA (twofold molar excess of oligonucleotide) along with chloramphenicol acetyltransferase (CAT) RNA as an internal control. The mixture was injected into the nuclei of CV-1 cells (approximately 150 cells per sample) and 4 h later the cells were fixed and immunolabelled for TAg and CAT expression (Table 1). Without oligonucleotide, all of the injected cells coexpressed CAT and TAg, and both the C-5 propyne- and butyne-substituted dU oligonucleotides completely inhibited TAg, but not CAT, expression. In contrast, neither of the thiazole-containing dU oligonucleotides inhibited TAg expression. Based on our knowledge of the mechanism of action of these compounds, we predict that the propyne- and butyne-containing oligonucleotides were bound to the RNA long enough to allow RNase H cleavage of the duplex in the cell, which resulted in inhibition of translation. In contrast, the thiazole-containing oligonucleotides dissociated from the duplexes prior to RNase H cleavage. This result was not anticipated, based on the observations from *in vitro* dissociation kinetics which indicated that the thiazole oligonucleotides should stay bound to RNA longer than the other duplexes, especially given the 4 h time point for the assay. Thus, rapid dissociation of thiazole-containing dU oligonucleotide/RNA duplexes may correlate with decreased antisense activity of the oligonucleotides (Gutierrez et al 1997).

Parameters that contribute to the activity of antisense oligonucleotides are complicated. In this study, we demonstrated that affinity, RNase H cleavage kinetics and intracellular dissociation all contribute to the activity of an oligonucleotide. In addition, we have previously shown that RNA secondary structure in the cell can also affect duplex formation and ultimately antisense activity. In this study, four oligonucleotides with comparable binding abilities were evaluated head-to-head in a series of assays. Clearly, the C-5 propyne-substituted oligonucleotides were the most active. The reason for the fourfold loss in activity between the C-5 propyne- and C-5 butyne-containing dU oligonucleotides is currently not understood, since none of the assays were predictive for the decreased activity of the C-5 butyne oligonucleotide. Intracellular dissociation of thiazole-containing dU oligonucleotide/RNA duplexes in the cell appeared to be largely responsible for the decreased antisense activity of the oligonucleotides. Similar factors account for the decreased activity of oligonucleotides that inhibit gene expression by a steric block mechanism and for the decreased activity of phosphorothioate oligonucleotides that contain unmodified bases (Moulds et al 1995).

Antisense activity using cytofectins

Recently there has been significant interest in the development of cationic lipids for the delivery of nucleic acids to a wide variety of cells. There are now many cytofectin agents commercially available for nucleic acid transfections from a wide range of vendors. One

TABLE 1 Biophysical data and biological activity of C-5-substituted oligodeoxynucleotides[a]

Oligodeoxynucleotide	T_m^b (°C)	K_a^c per s (rel)	K_d^d per s (rel)	K_d^e (K_a/K_d) (rel)	RNase H[f] (rel)	Pre-form[g]	IC_{50}^h (μM)
C-5 propyne	66.5	5.7×10^5 (1.0)	1.13×10^{-3} (1.0)	5.0×10^8 (1.0)	32 ± 7 (1.0)	100	0.25
C-5 butyne	66.0	7.7×10^5 (1.3)	1.08×10^{-3} (1.0)	7.1×10^8 (1.4)	27 ± 5 (0.84)	100	1.00
C-5 thiazole	67.5	5.1×10^5 (0.9)	4.44×10^{-4} (0.4)	1.1×10^9 (2.2)	14 ± 3 (0.44)	0	5.00
C-5 dimethylthiazole	69.0	1.3×10^6 (0.4)	4.83×10^{-4} (0.4)	2.7×10^9 (5.4)	5 ± 3 (0.16)	0	5.00

[a] Oligodeoxynucleotide = 5'CUUCAUUUUUUCUUC 3', where C = 5-methyl-2'-deoxycytidine and A = 2'-deoxyadenosine. All contain phosphorothioate linkages.
[b] T_m values were assessed in 140 mM KCl, 5 mM Na_2HPO_4, pH 7.2 at 260 nm, and the final concentration of oligodeoxynucleotide was about 2 μM (RNA concentration = about 2 μM).
[c] Pseudo first-order association rate constants for antisense oligodeoxynucleotides to RNA.
[d] The dissociation rate constants of antisense oligodeoxynucleotides at 37°C.
[e] The equilibrium binding constants for the oligodeoxynucleotides to RNA were determined from the quotient of K_a and K_d.
[f] The per cent of cleavage of RNA by RNase H in HeLa cell nuclear extract after 10 min at 37°C.
[g] The per cent inhibition of SV40 TAg translation in African Green Monkey CV-1 kidney cells by antisense oligodeoxynucleotides pre-bound to 5' capped, unspliced TAg RNA.
[h] The concentration of antisense oligodeoxynucleotide necessary for 50% inhibition of SV40 TAg after nuclear microinjection.
Relative values (rel) in parentheses are based on the values of the propyne oligodeoxynucleotide 5.
Reprinted with permission from Gutierrez et al (1997). Copyright (1997) American Chemical Society.

of the first cytofectins to be described was Lipofectin (Life Technologies, Gaithersburg, MD) which is a 1:1 (w/w) formulation of the cationic lipid DOTMA {N-[1-(2,3-dioleoyloxy)propyl]N,N,N-trimethylammonium chloride} and DOPE (dioleoyphosphatidylethanolamine) (Felgner et al 1987), and it has been used to deliver oligonucleotides to cells resulting in specific antisense gene inhibition (reviewed in Wagner 1994). Other examples of commercially available cytofectins include: LipofectAce (a 1:2.5 [w/w] formulation of dimethyl dioctadecylammonium bromide and DOPE; Life Technologies); LipofectAmine (a 3:1 [w/w] formulation of 2,3-dioleyloxy-N-[2(sperminecarboxamido)ethyl]-N,N-dimethyl-1-propanaminiumtrifluoroacetate and DOPE; Life Technologies); and the Transfectam reagent 5-carboxyspermylglycinedioctadecylamide; Promega Corp., Madison, WI).

In general, the antisense activity of an oligonucleotide can be realized using cytofectins to deliver oligonucleotides to a wide variety of cell lines. For antisense oligonucleotides that have been carefully selected to show potent activity, low nanomolar concentrations of oligonucleotides in cell media can cause significant (eight-to 10-fold) reductions in targeted RNA and protein levels in the cell. Most of the commercial cytofectin agents suffer from the facts that they need to be used in the absence of serum in the media, they can be toxic to cells and they are only active against a small subset of cell types (Lewis et al 1996).

Structure–activity relationships to identify cationic lipid formulations with improved properties

Although many researchers have had success delivering oligonucleotides to cells using the agents above, an improved cationic lipid, called GS 2888 cytofectin, has been developed for oligonucleotides that has several advantages (Fig. 1) (Lewis et al 1996). These include: (i) ease of use; (ii) use in the presence or absence of serum; (iii) delivery of oligonucleotides to a wide variety of cells; (iv) reproducible and uniform delivery of oligonucleotides to cells; and (v) relative non-toxicity to cells (Fig. 3). A detailed report describing the synthesis, formulation and use of the GS 2888 cytofectin has just been published and can be referred to for methods and properties of the agent (Lewis et al 1996).

The GS 2888 lipid is structurally derived from the Transfectam reagent, 5-carboxyspermylglycinedioctadecylamide, in which the 5-carboxyspermyl group was replaced with t-butoxycarbonyl (BOC)-arginine and the dioctadecylamide group was replaced with dimyristylamide. Based on observations from previous studies of cationic lipid structure–activity relationships, we evaluated dialkyl chain length using the t-BOC-arginine head group and found the relative order of effectiveness was C14 > C18 >> C12 when the cationic lipid was complexed in a 1:1 molar ratio with DOPE, formulated as a liposome and tested for oligonucleotide delivery properties (Lewis et al 1996). A similar relationship was previously found for derivatives of DOTMA (Felgner et al 1994). The C14 chain length may cause

0% FBS   **10% FBS**

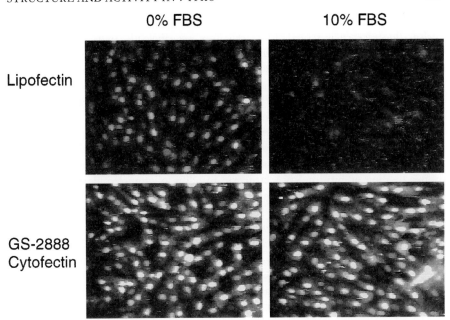

FIG. 3. Uptake of a 5' fluorescein-labelled 15 mer C-5 propyne phosphorothioate oligonucleotide into Rat-2 cells using Lipofectin (Life Technologies, Gaithersburg, MD) or the GS 2888 cytofectin. Rat-2 cells were incubated with either Lipofectin (10 µg/ml) or GS 2888 cytofectin (2.5 µg/ml) and fluorescent oligodeoxynucleotide (250 nM) for 24 h in Dulbecco's modified Eagle's medium with or without 10% fetal bovine serum (FBS). Confocal images (Noran Instruments, Middleton, WI) were captured (× 16 Plan-neofluar lens; Zeiss Axiovert 10 microscope [Zeiss, San Leandro, CA]) at identical settings. Images show fields of live cells in which the fluorescent oligonucleotide was delivered to their nuclei. Reprinted with permission from Lewis et al (1996). Copyright (1996) National Academy of Sciences.

decreased bilayer stiffness and phase transition temperatures compared to C18 and it is predicted to lead to more favourable fusion with endosomal membranes. Cellular uptake studies showed that the mono-BOC arginine cationic lipid was more efficient at delivering oligonucleotides to cells than the arginine without BOC (Lewis et al 1996). This was likely related to the relative pK_a of the polar head group. The guanidinium of arginine has a pK_a close to 9.0 and would be fully protonated at physiological pH. In contrast, the mono-BOC guanidinium group is expected to have a lower pK_a, close to 7. This observation has prompted further studies that evaluate head groups with different pK_as.

Methods for performing gene inhibition in cell culture

To determine the optimal concentration of GS 2888 cytofectin to use for a cell line, a dose titration of the lipid is performed, from 1.0 µg/ml to 20.0 µg/ml, using a 5' fluorescein-labelled oligonucleotide (250 nM) and the percentage of cells displaying nuclear fluorescence is scored. Most cell lines are efficiently transfected with the fluorescent oligonucleotide between 1.0 and 5.0 μg/ml of GS 2888 cytofectin. Using CV-1 cells, 95–100% of the cells show intense nuclear staining after 4–6 h of incubation with the lipid–oligonucleotide mix at the optimal concentration of GS 2888 cytofectin (2.5 μg/ml) (Lewis et al 1996). This assay is used in all of our antisense experiments to assure that efficient delivery is accomplished.

Selection of antisense oligodeoxynucleotides is not a straightforward process. Many researchers use nuclease-resistant phosphorothioate oligodeoxynucleotides targeted to the AUG initiation codon of a message. In our experience, these sites may not offer the proper binding affinity or RNA structure to allow oligonucleotide binding. The number of sequences to screen and the length of oligodeoxynucleotide to use is dependent on the modification of the oligodeoxynucleotide. Unfortunately, computer programs that use algorithms to predict RNA folding do a poor job of predicting RNA structure in the cellular environment. Thus, the inability to predict targetable regions of an RNA likely adds to the empirical nature of selecting antisense oligodeoxynucleotides (Flanagan & Wagner 1997). Additionally, the binding affinity of an oligodeoxynucleotide for its target is important, but it may be difficult to predict, owing to the presence of unwinding activities in cells that accelerate oligodeoxynucleotide/RNA duplex dissociation (Moulds et al 1995). It has been shown, however, that any region of the RNA can theoretically be targeted using antisense oligodeoxynucleotides (5' and 3' untranslated regions, AUG initiation codon, splice junctions, introns and coding sequences) provided that RNase H-competent oligodeoxynucleotides are used (Wagner et al 1993, Moulds et al 1995). Using 20 nucleotide (nt)-long phosphorothioate oligonucleotides without modified bases, at least 10–20 sequences should be screened to identify an active inhibitor. C-5 propyne oligonucleotides, in our studies, show higher potencies than oligonucleotides lacking the modification and active sequences are simpler to identify. The rule of thumb for selecting oligodeoxynucleotide sequences using C-5 propyne phosphorothioate oligodeoxynucleotides is straightforward: select six 15 nt antisense sequences that are complementary to any region of the target RNA and contain 50–80% pyrimidine content (to utilize the C-5 propynyl pyrimidines). Our experience has been that at least two to four of the six sequences will show potent activity (Fig. 4).

Elucidation of gene function using C-5 propyne oligonucleotides and GS 2888 cytofectin

C-5 propyne oligonucleotides have been used in several studies to evaluate the function of genes in cells. Predominantly, our own work has focused in the area of cell cycle proteins that may be involved in hyperproliferative disorders, particularly cancer. Several cell cycle proteins have been successfully targeted with C-5 propyne-modified oligonucleotides including: p34[cdc2], a serine/threonine kinase (Flanagan et al 1996, Flanagan & Wagner 1997); cyclin B1, which is expressed at the G2/M phase

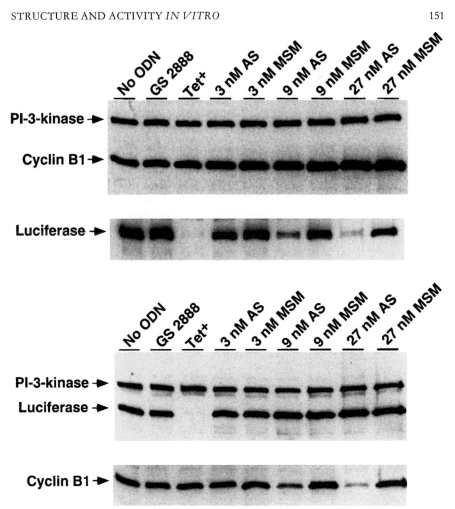

FIG. 4.   Antisense inhibition of luciferase using C-5 propyne oligodeoxynucleotides (ODNs).
Upper: HeLa X1/5 cells were incubated with either antisense (AS) or mismatch (MSM)
oligodeoxynucleotides targeted to luciferase (GS 2427 and GS 2631, respectively; 3, 9, and
27 nM, 24 h) together with the GS 2888 cytofectin (2.5 mg/ml) in Dulbecco's modified Eagle's
medium + 10% fetal bovine serum. Cell extracts were prepared and analysed by immunoblotting.
Since luciferase and cyclin B run at approximately the same molecular weight on the 7.5% SDS-
PAGE gel, side-by-side gels were subjected to electrophoresis using identical amounts of protein
per lane and immunoblotted using anti-PI-3 kinase, anti-luciferase and anti-cyclin B antibodies.
Since luciferase is negatively regulated by tetracycline in these cells, tetracycline (1 mg/ml) was
added to the media of HeLa X1/5 cells at the time of oligodeoxynucleotide transfection as a
positive control. Lower: antisense inhibition of cyclin B using C-5 propyne sense
oligodeoxynucleotides. HeLa X1/5 cells were incubated with either antisense or mismatch
oligodeoxynucleotides targeted to cyclin B (GS 3207 and GS 3319, respectively) in an
otherwise identical experiment to the upper panel. Each of the above experiments were
repeated in duplicate and similar results were obtained. Reprinted with permission from Lewis
et al (1996). Copyright (1996) National Academy of Sciences USA.

of the cell cycle (Fig. 4) (Flanagan et al 1996, Lewis et al 1996); and p27^{kip1}, a universal cyclin-dependent kinase inhibitor (CKI) (Coats et al 1996, St Croix et al 1996).

Many cell cycle proteins, including p34^{cdc2}, are up-regulated in cancer cells. p34^{cdc2} is tightly regulated during the normal cell cycle by phosphorylation, dephosphorylation and binding of cyclin B. It has been well documented that p34^{cdc2} is critical for cell cycle progression since dominant negative mutants of p34^{cdc2} inhibit cell cycle progression in the G2/M phase of the cell cycle (van den Heuvel & Harlow 1993) and temperature-sensitive mutants of p34^{cdc2} do the same when cells expressing the proteins are grown at the non-permissive temperature (Th'ng et al 1990). C-5 propyne antisense oligonucleotides targeted to p34^{cdc2} were developed in our laboratory (Flanagan et al 1996). These oligonucleotides showed mismatch-sensitive inhibition of p34^{cdc2} using Western blot analysis of normal human dermal fibroblast cells treated for 24 h with antisense oligonucleotides and GS 2888 cytofectin. Cell cycle analysis showed that cells treated with 1 nM antisense oligonucleotide showed a dramatic G2/M arrest, whereas a mismatch oligonucleotide showed no effect. p34^{cdc2} is up-regulated in breast carcinoma cells. Treatment of MCF-7 cells, a human breast carcinoma cell line, showed inhibition of p34^{cdc2} following antisense treatment, but had little effect on inhibition of the cell cycle. This result suggested that cancer cells may have redundant cell cycle pathways that compensate for p34^{cdc2} inhibition (Flanagan et al 1996).

p27^{kip1} is another cell cycle protein that is up-regulated in tumours (St Croix et al 1996). In this case, p27^{kip1} acts as a cell cycle inhibitor protein that may be utilized by tumour cells to regulate their growth. Some tumour cell lines are resistant to chemotherapeutics when grown as three-dimensional spheroid cultures, and the chemoresistance has been correlated with overexpression of p27^{kip1} (St Croix et al 1996). Antisense inhibitors of p27^{kip1} were developed using C-5 propyne oligonucleotides. These oligonucleotides were previously shown to inhibit p27^{kip1} expression in embryonic murine BALB/3T3 cells when delivered with GS 2888 cytofectin, and to promote cell growth when the cells were deprived of growth factors (Coats et al 1996). In the three-dimensional spheroid tumour cultures, treatment with p27^{kip1} antisense oligonucleotides using GS 2888 cytofectin caused a reduction in p27^{kip1} protein levels, relative to a mismatch control, made cells enter S phase, and increased the sensitivity of the cells to cytotoxic agents and gamma radiation. This study points to p27^{kip1} as a potential target for the treatment of chemoresistant, solid tumours (St Croix et al 1996).

Conclusions

During the past 18 years, there has been tremendous developmental effort aimed at improving antisense oligonucleotides and relatively few modifications have emerged. Recent technological advances — (i) the incorporation of nuclease-resistant backbone modifications into oligonucleotides; (ii) the discovery that C-5 propyne-modified oligonucleotides have enhanced affinity for their RNA target; and (iii) the development of serum-resistant cationic lipids to deliver oligonucleotides efficiently to cells in tissue culture — have solved many of the problems that have limited the use of antisense oligonucleotides in biological systems. C-5 propyne-modified antisense oligonucleotides can now be used as routine biological tools to validate therapeutic targets, elucidate the roles of proteins in complex biological pathways and, potentially, to treat human diseases.

Acknowledgements

I wish to thank M. Matteucci, B. Froehler, M. Flanagan and the entire Code Blocker team, past and present, for their many contributions and helpful discussions.

References

Bonham MA, Brown S, Boyd AL et al 1995 An assessment of the antisense properties of RNase H-competent and steric-blocking oligomers. Nucleic Acids Res 23:1197–1203

Coats S, Flanagan WM, Nourse J, Roberts JM 1996 Requirement of p27[kip1] for restriction point control of the fibroblast cell cycle. Science 272:877–880

Felgner PL, Gadek TR, Holm M et al 1987 Lipection: a highly efficient, lipid-mediated DNA-transfection procedure. Proc Natl Acad Sci USA 84:7413–7417

Felgner JH, Kumar R, Sridhar CN et al 1994 Enhanced gene delivery and mechanism studies with a novel series of cationic lipid formulations. J Biol Chem 269:2550–2561

Flanagan WM, Wagner RW 1997 Potent and selective gene inhibition using antisense oligodeoxynucleotides. Mol Cell Biochem, in press

Flanagan WM, Su L, Wagner RW 1996 Elucidation of gene function using C-5 propyne antisense oligonucleotides. Nat Biotechnol 14:1141–1145

Gutierrez AJ, Matteucci MD, Grant D, Matsumura S, Wagner RW, Froehler BC 1997 Antisense gene inhibition by C-5 substituted deoxyuridine containing oligodeoxynucleotides. Biochemistry 36:743–748

Hanvey JC, Peffer NJ, Bisi JE et al 1992 Antisense and antigene properties of peptide nucleic acids. Science 258:1481–1485

Lewis JG, Lin K-Y, Kothavale A et al 1996 A serum-resistant cytofectin for cellular delivery of antisense oligodeoxynucleotides and plasmid DNA. Proc Natl Acad USA 93:3176–3181

Matteucci MD, Wagner RW 1996 In pursuit of antisense. Nature (suppl) 384:20–22

Milligan JF, Matteucci MD, Martin JC 1993 Current concepts in antisense drug design. J Med Chem 36:1923–1937

Moulds C, Lewis JG, Froehler BC et al 1995 Site and mechanism of antisense inhibition by C-5 propyne oligonucleotides. Biochemistry 34:5044–5053

St Croix B, Florenes VA, Rak JW et al 1996 Impact of the cyclin dependent kinase inhibitor p27[kip1] on adhesion-dependent resistance of tumor cells to anticancer agents. Nat Med 2:1204–1210

Stein CA, Cheng Y-C 1993 Antisense oligonucleotides as therapeutic agents — is the bullet really magical? Science 261:1004–1012

Th'ng JPH, Wright PS, Hamaguchi J et al 1990 The FT210 cell line is a mouse G2 phase mutant with a temperature-sensitive CDC2 gene product. Cell 63:313–324

van den Heuvel S, Harlow E 1993 Distinct roles for cyclin-dependent kinases in cell cycle control. Science 262:2050–2054

Wagner RW 1994 Gene inhibition using antisense oligodeoxynucleotides. Nature 372:333–335

Wagner RW, Matteucci MD, Lewis JG, Gutierrez AJ, Moulds C, Froehler BC 1993
    Antisense gene inhibition by oligonucleotides containing C-5 propyne pyrimidines.
    Science 260:1510–1513
Wagner RW, Matteucci MD, Grant D, Huang T, Froehler BC 1996 Potent and selective
    inhibition of gene expression by an antisense heptanucleotide. Nat Biotechnol 14:840–844

DISCUSSION

Lebleu: J. P. Behr proposed that one of the key issues is that cationic lipids or polyethyleneimines act as a 'sponge' for endosomal protons (Boussif et al 1995). As a result, endosomes become destabilized. Have you observed anything like this with your cytofectins?

Wagner: We don't observe a general endosome destabilization. We have introduced a tagged dextran, where a fluorescein oligonucleotide is complexed with cationic lipid and Texas Red dextran, into the cell, and we observed that the dextran traffics into lysosomes. In contrast, the oligonucleotide complexed with cationic lipid is endocytosed into the cytoplasm and then into the nucleus.

Krieg: Is it a charged dextran or a neutral dextran?

Wagner: These were neutral dextrans.

Akhtar: Do you see any evidence for the by-passing of endocytosis with cationic lipid complexes simply by enhanced membrane permeation?

Wagner: No.

Crooke: Have you studied the metabolism of thiazole in cells?

Wagner: We don't know much about what happens to it in cells.

Crooke: What extract did you use for your RNase H studies?

Wagner: It was a HeLa cell extract.

Crooke: So there's probably not much point in doing the detailed kinetic studies. Where do you expect the thiazole to be located? In the major groove or the minor groove?

Wagner: Probably in the major groove.

Crooke: We showed with *Eschericia coli* RNase H that both major and minor groove modifications dramatically altered RNase H activity, depending on the nature of the modification, so it's not surprising that you observed a significant reduction in activity.

Wagner: We looked at the relative rates of RNase H cleavage using an oligonucleotide targeted to the 5′ leader sequence of T antigen. We found that this longer sequence which contained the C-5 thiazole dU modification had the same RNase H cleavage kinetics as the C-5 propyne dU control oligonucleotide, but we still saw the 20-fold decrease in activity for the C-5 thiazole oligonucleotide.

Crooke: Have you looked at whether bacterial helicases can unwind duplexes? Because if your notion is correct it will drive us away from looking at certain chemical modifications that might enhance activity.

Wagner: We haven't. We see a similar effect in some of the steric-blocking agents that we've used. If we preform a 2'-O-allyl C-5 propyne diester oligonucleotide that has extremely high RNA-binding affinity for RNA at the 5' untranslated cap site of T antigen, it also gets stripped off following microinjection into cell nuclei. We can show that this molecule has activity in the cell because it is active against plasmid-derived expression of T antigen, although it has 20-fold decreased activity compared to a C-5 propyne phosphorothioate oligonucleotide.

Crooke: But that's based on preforming the complex and then performing a cell assay, rather than on an isolated enzyme.

Wagner: Yes, but we observe the same result in an *in vitro* translation system.

Cohen: Since you're delivering your oligonucleotides in these improved liposomes and you're also forming RNA duplexes prior to the cell assays, do you see any improvements in activity with a phosphodiester backbone as opposed to a phosphorothioate one?

Wagner: We haven't done the experiment. I would expect that you would see decreased activity using an unmodified oligonucleotide.

Lebleu: When you use optimal conditions for oligonucleotide delivery, do you know the final concentration of the oligonucleotide in the cell?

Wagner: We can only estimate this based on the activity relative to microinjected oligonucleotides. We can definitely say that the cationic lipids concentrate oligonucleotides from the media into the cell. One of the interesting things is that for all the modifications we've looked at to date we have never been able to crack the 50 nM IC_{50} barrier.

Pieken: Your lipid has two well-defined cationic centres. What is known about the lipid binding? Are there dedicated lipid–oligonucleotide complexes or is there a degenerative absorption of oligonucleotides on the particle surface?

Wagner: It's the latter case. We mix the cationic lipid and the fusogenic lipid together in a 2 : 1 molar ratio. We freeze–thaw the mixture, sometimes we size it and sometimes we don't but it's more or less a liposomal preparation, and then we add that directly to the oligonucleotide and introduce it into the cells.

Gewirtz: Have you looked at this transfecting agent in other cell types?

Wagner: Yes we have, although it tends not to work on some, such as haematopoietic cells and T cells.

Agrawal: Do the oligonucleotides with propyne base modifications have improved nuclease stability?

Wagner: No.

Agrawal: Do you have any information on the safety and pharmacokinetic profiles of these compounds *in vivo*?

Wagner: We have done some limited studies and found that the C-5 propyne-modified oligonucleotides are more toxic than phosphorothioate compounds. Based on their activity in cell culture, we hope that they will be more potent *in vivo*, thus giving a better therapeutic index than phosphorothioates.

Agrawal: But your *in vitro* studies involved doing microinjections, and these conditions are artificial and not practical for common use.

Wagner: All the studies that we did with the cationic lipids, such as those described for p34^{cdc2} kinase and cyclin B, did not involve microinjections.

Caruthers: You mentioned that they were more toxic. Is this because of the breakdown products?

Wagner: It's difficult to say. Liver toxicity is likely. If the monomer is injected the toxicity is not observed, but this could also be due to the pharmacokinetics of the monomer versus the oligonucleotide.

Caruthers: Is the toxicity also dependent on the length of the oligonucleotide?

Wagner: Yes. Shorter oligonucleotides show greatly decreased toxicity.

Gewirtz: The presumed mechanism for the p27 antisense oligonucleotide is to get cells to cycle and thus increase their chemosensitivity. Did you do any experiments with non-cell cycle-activating chemotherapeutic agents?

Wagner: We have looked at cyclophosphamide and carboplatin.

Crooke: Platinum works both in cycling and non-cycling cells, whereas cyclophosphamide is not typically thought of as being a cell cycle-dependent agent.

Gewirtz: One point that I am concerned about, as a clinician, is getting cells to cycle in the presence of a chemotherapeutic agent because of the risk of mutagenesis, which may cause adverse effects five, 10 or 15 years later. An example of this is the incidence of secondary leukaemias in women who have been treated for breast carcinomas and ovarian cancer. Also, increased cell cycle activity could well increase acute toxicity.

Stein: In this particular case the patients would be lucky to survive for five, 10 or 15 years, so I wouldn't worry about the chronic toxicities at this stage. I also wanted to mention that in your cell cycle experiments there were no hypodiploid populations, so I was wondering whether you had any evidence of cellular apoptosis?

Wagner: Not in these cases.

Crooke: Another problem is achieving synchronization. It is possible that this has failed in the past because the therapeutic index of the compounds being used to try synchronize populations of cells are too toxic, although in my opinion the reasons are more complicated than this because the cells are extremely heterogeneous so it is difficult to get enough of the cells into a particular phase of the cell cycle.

Krieg: I would like to bring up the question of the use of these differential oligonucleotide backbones for gene validation purposes, i.e. to study the biological function of newly cloned genes. This is related to the frequency of the targetable sequences on genes. Some people have stated that about 80% of the time, a propyne antisense oligonucleotide will greatly reduce the target RNA levels, and that the success rate for phosphorothioate antisense oligonucleotides is substantially lower than this.

Crooke: That has not been our experience. We have looked at many targets with many different chemistries and we have found that higher affinity compounds have increased potency. Also, more sites are available as more structure can be invaded and there is a higher hit rate, but it is nowhere near 80%.

Wagner: Typically, we take six sequences and of those we find two that are in the potency range for showing antisense effects, so that's 30%.

Matteucci: That is still higher than a regular phosphorothioate compound. One of the things that may be biasing this is that by using oligomers containing a high content of propynyl uracil we are targeting adenosine-rich regions of the RNA, which have little secondary structure.

Monia: You mentioned the use of the cytofectin reagent as a laboratory tool. Have you tested this *in vivo* and do you envisage any *in vivo* applications for cationic lipids?

Wagner: We've attempted to test them *in vivo,* but we haven't yet found the optimal conditions for them to work effectively. In terms of *in vivo* applications, in general one wouldn't want to use cationic lipids because they would make extremely complicated drugs.

Crooke: Unless you wanted to use them in the reticulo-endothelial system.

Agrawal: Is liver toxicity caused only by propyne oligonucleotides or did you observe it when you used other base-modified oligonucleotides?

Crooke: Not all base-modified oligonucleotides cause liver toxicity, at least not all the ones we've tested.

Gait: Now that higher affinity oligonucleotides are available, is it possible to use shorter oligonucleotides? You have already published one example of this, but is it likely to be a general phenomenon?

Wagner: It is possible. I don't have many other examples so I would not like to say that it will be a general phenomenon.

Southern: The reason is probably that the sequence specificity is determined by structural folding, i.e. certain regions are exposed in subtargets, and the same sequence may be present in other targets but not accessible for hybridization.

Caruthers: Do smaller oligonucleotides have the same kinetic on-rates?

Iversen: I don't understand why anyone would want to use shorter oligonucleotides of the order of 7 nt because that sequence may occur 10 000 times in the genome. Wouldn't this result in non-specific oligonucleotides?

Wagner: The point is that one gains specificity by targeting a unique structure of the RNA, thus the flanking sequences contribute to specificity.

Crooke: But whether we want them to or not, these oligonucleotides are identifying structure as well as sequence. Therefore, it's difficult to predict the level of hybridization specificity based strictly on the mathematics of hybridization.

Reference

Boussif O, Lezouala'h F, Zanta MA et al 1995 A versatile vector for gene and oligonucleotide transfer into cells in culture and *in vivo*: polyethyleneimine. Proc Natl Acad Sci USA 92:7297–7301

Progress in antisense therapeutics discovery and development

Stanley T. Crooke

Isis Pharmaceuticals Inc., Carlsbad Research Center, 2292 Faraday Avenue, Carlsbad, CA 92008, USA

Abstract. Progress in the discovery and development of antisense therapeutics continues at a rapid pace. Detailed understanding of the pharmacokinetic, toxicological and pharmacological properties of phosphorothioate oligonucleotides is now available. Further, an ever-increasing body of information is available about new and improved chemical classes of antisense drugs that are progressing toward clinical trials.

1997 Oligonucleotides as therapeutic agents. Wiley, Chichester (Ciba Foundation Symposium 209) p 158–168

During the past few years, interest in developing antisense technology and in exploiting it for therapeutic purposes has been intense (Crooke et al 1995, 1996). Although progress has been gratifyingly rapid, the technology remains in its infancy and the questions that remain to be answered still outnumber the questions for which there are answers. Appropriately, considerable debate continues about the breadth of the utility of the approach and about the type of data required to 'prove that a drug works through an antisense mechanism' (Crooke 1995b). The objectives of this chapter are to provide a summary of recent progress, to assess the status of the technology and to place the technology in the pharmacological context in which it is best understood.

Proof of mechanism

Until more is understood about how antisense drugs work, it is essential to demonstrate effects that are consistent with an antisense mechanism. For RNase H-activating oligonucleotides, Northern blot analysis showing selective loss of the target RNA is the best choice and many laboratories are publishing reports *in vitro* and *in vivo* of such activities (Crooke et al 1994, 1995a, 1996). Ideally, a demonstration that closely related isotypes are unaffected should be included. In brief, then, for proof of mechanism, the following steps are recommended (Crooke 1995b):

(1)   perform careful dose–response curves *in vitro* using several cell lines and methods of *in vitro* delivery;

(2)   correlate the rank order potency *in vivo* with that observed *in vitro* after thorough dose–response curves are generated *in vivo*;

(3)   perform careful time courses before drawing conclusions about potency;

(4)   directly demonstrate proposed mechanism of action by measuring the target RNA and/or protein;

(5)   evaluate specificity and therapeutic indices via studies on closely related isotypes and with appropriate toxicological studies;

(6)   perform sufficient pharmacokinetics to define rational dosing schedules for pharmacological studies; and

(7)   when control oligonucleotides display surprising activities, determine the mechanisms involved.

Molecular mechanisms

Figure 1 shows the process by which genomic information is converted to proteins. In effect, antisense drugs are designed to inhibit the intermediary metabolism of RNA. Figure 2 shows a number of potential mechanisms of action that have been

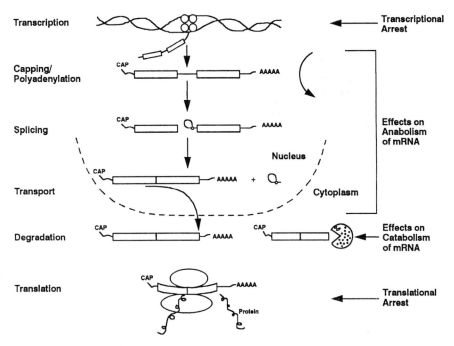

FIG. 1.   RNA processing.

demonstrated. To date, the best characterized and most potent mechanism is the activation of RNase H.

Characteristics of phosphorothioate oligonucleotides

Of the first-generation antisense drugs, the chemical class that has proven of most value is the phosphorothioate class. In this chemical class, a non-bridging oxygen in the phosphate group is replaced by a sulfur. This creates a chiral centre at each phosphate, enhances nuclease resistance and results in a significant increase in low affinity interactions with many proteins (Crooke et al 1995, Crooke & Bennett 1996).

The pharmacokinetic and toxicological properties of many phosphorothioate oligonucleotides have been characterized in multiple species and in humans (Cossum et al 1993, 1994, Crooke et al 1995). The pharmacokinetic properties of these drugs can be summarized as follows:

(1)   they are well absorbed from all parental sites;
(2)   they are distributed rapidly;
(3)   they bind to serum albumin;
(4)   they reside for prolonged periods of time in tissues;

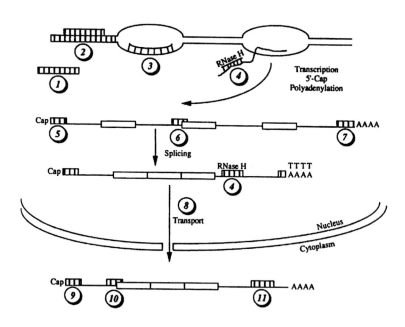

FIG. 2.   Sites of action for oligonucleotide drugs.

(5) they have a broad peripheral tissue distribution and they are not distributed in the brain;

(6) they are cleared by metabolism over a prolonged period ($\geqslant$ 50% over 10 days);

(7) this metabolism is consistent across species at doses of 0.5–3 mg/kg (humans probably metabolize slightly more slowly than humans);

(8) there are minimal differences between sequences; and

(9) they may have dose-dependent pharmacokinetics.

The data that support the notion that these drugs are localized intracellularly in various organs are derived from several sources: autoradiography of proximal convoluted tubule cells, keratinocytes and liver (parenchymal and Kupffer) cells; fluorescently tagged oligonucleotides of all the above cell types in addition to endothelial cells and other cells in the colon; immunohistochemistry with monoclonal antibodies that recognize intact oligonucleotides; and the isolation of liver cells and subcellular fractionation after systemic dosing.

Similarly, the toxicological properties of phosphorothioate oligonucleotides include the following:

(1) they have no *in vitro* cytotoxicity at concentrations of less than 100 mM;

(2) they are non-mutagenic;

(3) they have no significant subacute toxicities in rodents and monkeys after intravenous doses of up to 100 mg/kg;

(4) they do not demonstrate delayed-type hypersensitivity;

(5) single and multiple doses have been demonstrated to be safe in over 400 humans treated with ISIS 2105;

(6) high dose toxicity (> 80 mg/kg per day) in rodents is likely related to cytokine release;

(7) high doses (72 mg/kg per day) inhibit coagulation; and

(8) there are hypotensive effects in monkeys which may complement activation.

Figure 3 provides a sense of the therapeutic index of these drugs by comparing the dose required to achieve plasma concentrations associated with various adverse events with those at which pharmacological activity has been achieved. Given that doses as low as 0.06 mg/kg per day intravenously have been reported to be effective (Monia et al 1996) in some models, the therapeutic index would seem satisfactory for many indications.

Numerous studies have demonstrated *in vivo* activities of antisense phosphorothioate oligonucleotides in animals (Crooke & Bennett 1996, Crooke et al 1996). Further, a growing number of well-controlled trials have been reported that provide convincing evidence that the observed activities are due to an antisense mechanism. Finally, studies on two drugs, ISIS 2922 (fomivirsen) and ISIS 2302 in humans have yielded evidence of efficacy. ISIS 2922 is in Phase III trials for patients

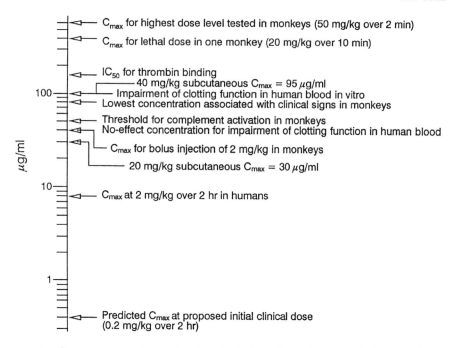

FIG. 3.   Plasma concentration and various biological effects of ISIS 2302. C_{max}, maximum plasma concentration.

with cytomegalovirus retinitis, and ISIS 2302 has recently completed a randomized placebo controlled Phase II trial in patients with Crohn's disease (Yacyshyn et al 1997).

Thus, the phosphorothioates have performed better than many might have expected. Nevertheless, they have numerous significant pharmacodynamic (they have a low affinity per nucleotide unit and they inhibit RNase H at high concentrations), pharmacokinetic (they have limited oral availability and blood–brain barrier penetration, and their pharmacokinetics are dose dependent) and toxicological (they cause the release of cytokines, promote clotting and may have complement-associated effects on blood pressure) limitations (Crooke et al 1995).

Medicinal chemistry

Figure 4 shows a dinucleotide and summarizes the novel modifications introduced and tested at Isis Pharmaceuticals Inc. in the past several years (Cook 1993). Based on these studies we are now confident that new chemical classes with improved potency, pharmacokinetic and toxicological properties will be available and will shortly reach the clinic.

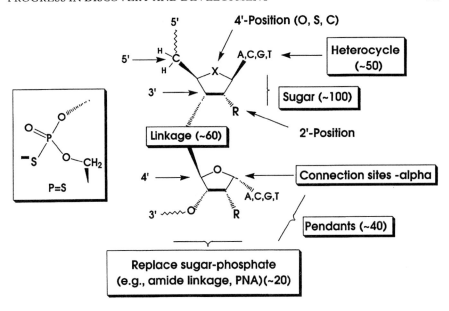

FIG. 4.   Isis oligonucleotide modifications showing sites and numbers of modifications that have been synthesized and tested. PNA, peptide nucleic acid; P = S, phosphorothioate.

Conclusions

Although many more questions about antisense technology remain to be answered, progress has continued to be gratifying. Clearly, as more is learned we will be in the position to perform more sophisticated studies and to understand more of the factors that determine whether an oligonucleotide actually works via antisense mechanisms. We should also have the opportunity to learn a great deal more about this class of drugs as additional studies are completed in humans.

Acknowledgements

I thank the many people at Ciba and Isis, as well as other collaborators, who contributed to this work. I also thank Donna Mussachia for her administrative assistance.

References

Cook PD 1993 Medicinal chemistry strategies for antisense research. In: Crooke ST, Lebleu B (eds) Antisense research and applications. CRC Press, Boca Raton, FL, p 149–187

Cossum PA, Sasmor H, Dellinger D et al 1994 Disposition of the ^{14}C-labeled phosphorothioate oligonucleotide ISIS 2105 after intravenous administration to rats. J Pharmacol Exp Ther 267:1181–1190

Cossum PA, Truong L, Owens SR, Markham PM, Shea JP, Crooke ST 1994 Pharmacokinetics of a ^{14}C-labeled phosphorothioate oligonucleotide ISIS 2105 after intradermal administration to rats. J Pharmacol Exp Ther 269:89–94

Crooke RM, Graham MJ, Cooke ME, Crooke ST 1995 *In vitro* pharmacokinetics of phosphorothioate antisense oligonucleotides. J Pharmacol Exp Ther 275:462–473

Crooke ST 1995a Therapeutic applications of oligonucleotides. RG Landes, Austin, TX

Crooke ST 1995b Editorial: delivery of oligonucleotides and polynucleotides. J Drug Target 3:185–190

Crooke ST, Bennett CF 1996 Progress in antisense oligonucleotide therapeutic. Annu Rev Pharmacol Toxicol 36:107–129

Crooke ST, Grillone LR, Tendolkar A et al 1994 A pharmacokinetic evaluation of ^{14}C-labeled afovirsen sodium in patients with genital warts. Clin Pharm Therap 56:641–646

Crooke ST, Graham MJ, Zuckerman JE et al 1996 Pharmacokinetic properties of several novel oligonucleotide analogs in mice. J Pharmacol Exp Ther 277:923–937

Monia BP, Johnston JF, Geiger T, Muller M, Fabbro D 1996 Antitumor activity of a phosphorothioate antisense oligodeoxynucleotide targeted against c-*raf* kinase. Nat Med 2:668–675

Yacyshyn B, Woloshuk, B, Yacyshyn MB et al 1997 Efficacy and safety of ISIS 2302 (ICAM-1 antisense oligonucleotide) treatment of steroid-dependent Crohn's disease. Gastroenterology 112:A1123

DISCUSSION

Caruthers: Do you have any idea how many compounds are currently in clinical studies?

Crooke: ISIS 2922 (fomiversen), ISIS 3521 and ISIS 5132 (CGP 69846A) are in clinical trials. ISIS 2105 was in clinical trials but we pulled it out. GEM 91 and GEM 132 are in clinical trials, the latter of which is being administered systemically. Also, the Bcl2 compound has been given to nine or 10 patients so far.

Fearon: There is also LR-3280 for the prevention of restenosis and LR-3001 for leukaemic bone marrow purging *ex vivo*.

Crooke: There is reasonable evidence that several of these in animals work only via an antisense mechanism. With systemic administration, the Hybridon and Isis experiences have been fairly similar. GEM 91 is a 25 mer and is given to AIDS patients. ISIS 3521 and ISIS 5132 are being tested in patients with solid tumours by giving them 3 mg/kg 2 h or 24 h infusions, and we have not observed any adverse effects. There are now over 100 patients who have been treated systemically with 3–4 mg/kg. The only adverse event that we experienced was the subcutaneously administered ISIS 2302, which resulted in lymphadenopathy in the draining lymph nodes.

Nicklin: The issue of oral bioavailability is an important one. It's fairly easy to interpret oral bioavailabilities of that level (10–20%) based on radiochemical quantitation with ^{35}S-labelled compounds. If one looks at specific analysis with a range of modifications it is difficult to draw any conclusions, other than that bioavailability estimates of < 1% are probably more realistic. Have you looked at the specific analysis of the SATE compound?

Crooke: The results will be available soon. One of the problems with the way we have been doing the oral bioavailability studies, which involve comparing tissues, is that it tends to magnify the error. In some of the new studies, using capillary gel electrophoresis, with methoxyethoxy we're seeing significantly more bioavailability than 1%.

Nicklin: We've done these analyses with capillary gel electrophoresis, and we've also used ^{14}C labels. In rats and monkeys it's significantly less than what was reported in the literature (Agrawal et al 1995). It's an important issue but we shouldn't get too carried away with it.

Crooke: A lot depends on how the oral bioavailability is analysed and there are many caveats. First of all, many of these studies have only been performed in rodents, and until we test these modifications in monkeys we don't know whether the rodent data will be relevant. Second, the analyses are not necessarily the best way to look at oral bioavailability because of variations in plasma distribution and tissues, although more recently these techniques have improved. There is also significant variation in the results from lab to lab.

Agrawal: Another point we need to consider is that in our oral adsorption studies in fasting animals we see reduced adsorption. Therefore, it is necessary to fast the animals overnight, but give them a normal diet before oral administration. The plasma concentration studies suggest that the bioavailability is less than 1%, but this may not be the correct way to calculate the bioavailability of oligonucleotides. We have demonstrated that after 24 h about 30–40% of the administered dose is distributed throughout various tissues (Agrawal et al 1995). Will this influence biological effectiveness? In recent studies we have administered end-modified oligonucleotides targeted to protein kinase A orally in tumour-bearing mice, and we observed that about 1.5% of the administered dose was present in the tumours. The efficacy of the orally administered oligonucleotide was similar to the intraperitoneal administered oligoncleotide, suggesting that a slow-delivery system was used. Chemical modifications of oligonucleotides have an impact on adsorption from the gastro-intestinal tract. For example, end-modified oligonucleotides with methyl-phosphonate linkages show increased adsorption. The site of adsorption is also essential; we tried large intestine administration in anaesthetized animals and found that even phosphorothioate oligonucleotides were adsorbed intact, suggesting a rectal delivery mechanism was operating.

Nicklin: The problem first has to be considered from a pharmacokinetic perspective rather than a pharmacodynamic perspective, which would complicate the issue. With essentially identical experiments, our data are still inconsistent with those of others.

Crooke: I suspect that there are enough differences in the experimental details to account for the differences. The important point is, at the level we're at, it is by no means guaranteed that we're going to achieve useful oral bioavailability.

Nicklin: After oral administration we can't detect any significant levels of intact compound using gel electrophoresis of the kidneys. However, if we administer the compound by the pulmonary route using radioactive isotopes we get a bioavailability of about 40%, and this is comparable (~ 45%) to results using capillary gel electrophoresis.

Crooke: That's a good point. We are also looking at pulmonary administration, although it also has some caveats. We have not used aerosols, although it should be possible to do this with these compounds.

Iversen: We have looked at the oral bioavailability of three phosphorothioates—a 15 mer (LR-3280), a 20 mer (P53) and a 24 mer (c-*myb*). The most efficient was the 24 mer; it was 10-fold more efficient than the 20 mer, and the 15 mer was somewhere in-between.

Crooke: In a situation in which the limiting event is intragut metabolism any sequence differences will give rise to significant differences in bioavailability. We need to develop analogues that are not metabolized in the gut in order to sort out what modifications increase absorption.

Agrawal: When we injected phosphorothioate or end-modified phosphorothioate oligonucleotides intravenously, 2–3% of the injected dose was detected in bile, suggesting that enterohepatic circulation had occurred. However, we did not observe faecal clearance to the same level, suggesting that reabsorption had occurred.

Lebleu: Immunohistochemistry is a useful tool to follow oligonucleotide distribution. What do your antibodies recognize?

Crooke: They are sequence-specific antibodies. We have raised two against ISIS 2105, one recognizes 10 nt at the 5′ end and the other 10–15 nt at the 3′ end.

Lebleu: Are they recognizing the same sequence in different chemistries?

Crooke: We haven't looked at whether these antibodies recognize a peptide nucleic acid sequence, although I would be surprised if they did. Within the phosphorothioates, however, the antibodies are sequence specific.

Wickstrom: I'm sure you want to emphasize that you use a carrier protein to obtain these antibodies. Valentine Vlassov has worked on this and has not found antibodies against free oligonucleotides. This is an important point because some people still believe that oligonucleotides are immunogenic.

Crooke: We tried to make antibodies to phosphorothioates in every way humanly possible and we couldn't. I assumed that most people would know that we conjugate the oligonucleotides to antigenic peptides.

Southern: Can you be so categorical as to say that there's no sequence-specific immunogenicity?

Crooke: All phosphorothioates are immunostimulatory, and there are sequence differences in immunogenicity, but there are also sequences that produce significant immunogenicity but don't fit that rule. There is sequence dependence within the phosphorothioate group for potency for immunogenicity.

Wickstrom: I would like to endorse the observations of Stanley Crooke and Sudhir Agrawal on intestinal uptake. Rudy Juliano and I did some experiments with ^{14}C-labelled methylphosphonates, phosphorothioates and phosphorodithioates. We didn't do whole-animal studies, rather isolated pieces of ileum, and we observed slow paracellular uptake, i.e. about 5% per hour. However, the fact that the small intestine is so long in humans, at least, means that this is perfectly adequate for getting the kinds of bioavailabilities that Sudhir Agrawal is seeing (Hughes et al 1995).

Nicklin: The viability of the epithelium in the everted sac preparation is about 2 min once it is removed from the animal. If you look at $CaCO^{-2}$ monolayers as a model, with all their problems, they would predict much less than 1% oral absorption.

Wagner: How viable are the liver cells in your experiments after extraction?

Crooke: We have shown that after extraction and collagenase treatment the cells have about 90% viability. We then plate out the hepatocytes using standard methods.

Wagner: Have you tried those experiments using fluorescently labelled oligonucleotides?

Crooke: The experiments I described were performed with unlabelled oligonucleotide using solid phase extraction and capillary gel electrophoresis. However, we have also looked at fluorescently labelled oligonucleotides and we observe the same results. Eric Markson is doing the sister experiments to our gel electrophoresis experiments using fluorescently labelled oligonucleotides. There are reasonable correlations between these results, but it's early days yet and I'm sure there will be imperfections in those correlations.

Nicklin: It would be interesting to see if the subcellular distribution experiments correlate with the protein-binding properties, i.e. where you see protein you also see oligonucleotide and where you do not see protein you do not see oligonucleotide.

Cohen: Several years ago Bruce Dolnick (personal communication 1990) told me about some of his unpublished results on sequence-specific uptake and distribution of oligonucleotides into cells. The differences that he saw led him to conclude that antisense techniques would never be really effective, if there are such radical differences in uptake, and antisense oligonucleotides will only work when they can selectively enter the cell.

Crooke: But there will be different dose–response curves depending on where you want the drug, the intrinsic properties of the chemical class and the sequence. These are early data and they need to be confirmed.

Akhtar: Can you improve the oligonucleotide delivery by pre-saturating the Kupffer cells with other drugs or particulates?

Crooke: We don't know. We expected to see the drug in the Kupffer cells and in the hepatocytes but it seems that, based on time course experiments, the drug is not travelling to the hepatocyte as a primary event, although this needs to be confirmed.

Akhtar: If the drug has to pass through the vasculature to the Kupffer cells and eventually into the hepatocytes, it is possible that some of the phosphorothioates, for example, would be susceptible to degradation along these transcellular pathways.

Crooke: But we know that the half-lives of typical phosphorothioate in the liver are in excess of 48 h.

Nicklin: It's fairly easy to block the hepatic uptake of oligonucleotides. What tends to happen is that they are redistributed to other tissues such as muscle rather than to other cell types within the liver.

Gait: We have talked very little about antivirals. Combination therapy often seems to be effective. Have you tried fomivirsen together with more conventional antivirals, such as ganciclovir?

Crooke: Yes. It's additive. It also doesn't interfere with the action of AZT (3'-azido, 3'-deoxythymidine). We have looked at ganciclovir-resistant lines and the drug was active against those. We've also spent some time trying to generate viral lines resistant to fomivirsin itself. Kevin Anderson has been working on this.

Gewirtz: Have you looked at the levels of γ-interferon in the intraocular compartments? You observed some degree of inflammation, so γ-interferon synthesis may have been induced and may have been responsible for some of the antiviral effects you observed.

Crooke: We have not measured it in HIV patients treated with this drug so I can't answer the question directly. In the monkey we don't observe significant levels of γ-interferon. On the other hand, I can't exclude the possibility that the inflammatory properties in the molecule are contributing to the inflammation that we're seeing. We will have the chance to exclude this possibility because the next compound in the ISIS 2922 series is a methoxyethoxy, which in the monkey is not inflammatory.

References

Agrawal S, Zhang X, Lu Z et al 1995 Absorption, tissue distribution and *in vivo* stability in rats of a hybrid antisense oligonucleotide following oral administration. Biochem Pharmacol 50:571–576

Hughes JA, Avrutskaya AV, Brouwer K, Wickstrom E, Juliano RL 1995 Radiolabeling of methylphosphonate and phosphorothioate oligonucleotides and evaluation of their transport in everted rat jejunum sacs. Pharmacol Res 12:817–824

Oligonucleotide therapeutics for human leukaemia

Alan M. Gewirtz

Departments of Pathology and Laboratory Medicine and Internal Medicine, University of Pennsylvania School of Medicine, Philadelphia, PA 19104, USA

Abstract. The concept of antisense oligonucleotide 'therapeutics' has generated a great deal of controversy. Questions abound regarding the mechanism of action of these compounds, their reliability and their ultimate utility. These problems are compounded by the 'hype', which has attended their development, and the inability of workers in this area to meet the expectations raised by its most zealous proponents. Nevertheless, it is worth pointing out that there have been some notable gene disruption successes with this technique that have stood up to rigorous scrutiny. Our own work with c-*myb* as a target is perhaps a reasonable example. Though much remains to be accomplished before antisense drugs are commonly, and usefully, employed in the clinic, it is important to remember what motivates their development. Gene-targeted drugs have the promise of exquisite specificity and the potential to do much good with little toxicity. Accordingly, antisense oligonucleotides can serve as a paradigm of rational drug development. For all these reasons then, we believe that efforts should be increased to decipher the mechanism of action of antisense oligodeoxynucleotides, and to learn how they may be successfully employed in the clinic.

1997 Oligonucleotides as therapeutic agents. Wiley, Chichester (Ciba Foundation Symposium 209) p 169–194

In recent years, a large number of extracellular growth factors that regulate haemopoietic cell development have been cloned, expressed as active proteins and utilized in the clinic. The intracellular events that are triggered when these growth factors interact with their receptors have also begun to be defined. Nevertheless, most of the molecular machinery that regulates blood cell development remains enigmatic and difficult to access. This situation particularly applies to normal blood cells because of the difficulty of applying modern molecular analytical techniques to the small numbers of cells that are ordinarily available for such investigations.

The problem of understanding gene function in haemopoietic cells has involved two main strategies. One involves infecting cells with a vector engineered to express the gene of interest at high levels (Clarke et al 1988, Liebermann & Hoffman-Liebermann 1989, Prochownik et al 1990). If the cell's phenotype/behaviour changes in the infected cell, but not a sham-transfected cell, one can tentatively attribute the change in phenotype to the newly expressed gene. Function may therefore be inferred. This approach, while potentially informative, has a number of drawbacks. First, it is not certain that the changes observed are directly related to the gene's function as such changes may be indirect. Second, the experiment is not physiological because overexpression of the gene may exaggerate its normal function or impart new functions by leading to an excess of the encoded protein. Finally, this approach is often limited to leukaemic cell lines because normal cells are often much more problematic in terms of both infection and expression of the vector.

An alternative experimental approach, and one which may yield more physiologically relevant data, is either to physically knock-out the target gene or to interfere with its function by perturbing the use of its mRNA. Homologous recombination remains the 'gold standard' for experiments of this type because this methodology physically destroys the gene of interest (Galli-Taliadoros et al 1995, Heyer & Kohli 1994, Morrow & Kucherlapati 1993, Osman et al 1994, Willnow & Herz 1994). However, because this strategy ultimately depends on selecting cells in which a rare cross-over event has occurred, it would appear to have limited therapeutic practicality. Ribozymes, antisense RNAs and antisense DNAs, all of which interfere with the use of the targeted mRNA, appear to have more immediate relevance from a therapeutic point of view. For these reasons, these approaches have become an increasingly popular strategy for exploring gene function in cells of virtually any type.

For the past several years we have been engaged in trying to develop an effective strategy of disrupting specific gene function with antisense oligodeoxynucleotides. We have also been actively engaged in attempting to utilize this strategy in the clinic. This latter pursuit has focused on finding appropriate gene targets that can be successfully targeted using an antisense approach and on developing scale-up methods, so that techniques developed in the laboratory can be applied in the clinic. It was our opinion that human leukaemias would be particularly amenable to this therapeutic strategy. They can be successfully manipulated *ex vivo*, the tumour is 'liquid' *in vivo* and therefore more likely to successfully take up oligodeoxynucleotides, and a great deal is known about their cell and molecular biology. The latter in particular facilitates choice of a gene target. Accordingly, if oligodeoxynucleotides are going to be developed as therapeutics, the haemopoietic system seems an ideal model system.

The c-Myb proto-oncogene

Of the genes that we have targeted for disruption using the antisense oligodeoxynucleotide strategy (Gewirtz & Calabretta 1988, Luger et al 1996, Ratajczak et al 1992a,c, Small et al 1994, Takeshita et al 1993) one that has been of particular interest to our laboratory, and one where therapeutically motivated disruptions are now in clinical trial, is c-*myb* (Lyon et al 1994). c-*myb* is the normal cellular homologue of v-*myb*, the transforming oncogene of the avian myeloblastosis virus (AMV) and avian leukaemia virus E26. It is a member of a family composed of at least two other highly homologous genes designated A-*myb* and B-*myb* (Nomura et al 1993). It is located on chromosome 6q in humans and its predominant transcript encodes a ∼75 kDa DNA-binding protein (Myb) which recognizes the core consensus sequence 5 PyAAC(G/Py)G 3′ (Biedenkapp et al 1988). Myb consists of three primary functional regions (Sakura et al 1989) (Fig. 1).

At the N-terminus is the DNA-binding domain. This region consists of three imperfect tandem repeats (R1, R2, R3) each consisting of 51–52 amino acids. Three perfectly conserved tryptophan residues are found in each repeat. Together they form a cluster in the hydrophobic core of the protein, which maintains the DNA-binding helix-turn-helix structure. The mid-portion of the protein contains an acidic transcription-activating domain. The DNA-binding portion of the protein is required for these transcriptional effects to be observed. The protein also contains a negative regulatory domain which has been localized to the C-terminus. Interestingly, the C-terminus is deleted in v-*myb* and this has been thought to contribute to v-*myb*'s transforming ability. Recently reported experiments have confirmed this hypothesis and have further demonstrated that N-terminal deletions give rise to a protein with even more potent transforming ability (Dini et al 1995). Deletions of both the N- and C-termini create a protein with the greatest transforming ability, and one which induces the formation of haemopoietic cells that are more primitive than those produced by N-terminal deletions alone (Dini et al 1995). These data suggest that the simultaneous loss of Myb's ability to bind DNA and interact with as yet unidentified proteins is a potent transforming stimulus. Nevertheless, this simple hypothesis is complicated by the observation that overexpression of the C-terminal portion of c-*myb* can also be oncogenic (Press et al 1994), whereas overexpression of the whole protein is not (Dini et al 1995). At the very least, one may conclude that sequestration of certain potential Myb-binding proteins may also be an oncogenic event.

Recently, a putative leucine zipper structure was described within the N-terminal portion of Myb's C-terminal domain (Kanei-Ishii et al 1992). Leucine zippers, such as those found in the transcription factors Jun, Fos and Myc, are thought to facilitate the protein–protein interactions that permit heterodimerization of DNA-binding

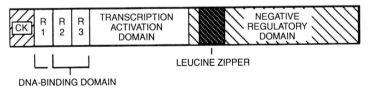

FIG. 1. Functional map of Myb.

proteins. Such dimerization is thought to play a key role in regulating the transcriptional activity of these factors. A Myb dimerizing binding partner has yet to be identified but Myb–Myb homodimerization, which likely occurs through its leucine zipper, does lead to loss of DNA binding and transactivation ability (Nomura et al 1993). Accordingly, one could reasonably postulate that Myb-driven transactivation and/or transformation might be regulated by the binding of additional protein partners in the leucine zipper domain (Kanei-Ishii et al 1992). Alternatively, loss of the ability of Myb to dimerize with a putative regulatory partner might also contribute, directly or indirectly, to cellular transformation and leukaemogenesis. Point mutations in the Myb negative regulatory domain might be one mechanism for bringing about such a loss (Kanei-Ishii et al 1992). Finally, interaction (not physical dimerization) with other nuclear-binding proteins, such as the CCAAT enhancer-binding protein (C/BEP) (Burk et al 1993) and the related myeloid nuclear factor NF-M (Ness et al 1993), may also regulate Myb's transactivation or repressor functions.

The above discussion suggests that c-*myb* might play a role in leukaemogenesis. Additional, albeit indirect, evidence also support this contention; for example, c-*myb* amplification in acute myelogenous leukaemia (AML) and overexpression in lymphomas associated with deletions on chromosome 6 have been reported (Barletta et al 1987). The mechanism whereby overexpressed c-*myb* might be leukaemogenic is uncertain but it points out the important difference of working with primary cells as opposed to cell lines. As noted above, it has been reported that overexpressing c-*myb* is not by itself leukaemogenic (Dini et al 1995) but this work was carried out in cell lines, which may give results that are valid only for the lines tested. As was also noted above, one could reasonably postulate that Myb-driven transformation might be regulated by the binding of additional protein partners in the leucine zipper domain (Kanei-Ishii et al 1992). Recent evidence demonstrating that Myb interacts with other nuclear-binding proteins, and that Myb's C-terminus may interact with a cellular inhibitor of transcription supports this hypothesis (Burk et al 1993, Vorbrueggen et al 1994). Other potential mechanisms might relate to Myb's ability to regulate haemopoietic cell proliferation (Gewirtz et al 1989), perhaps by its effects on important cell cycle genes including c-*myc* (Cogswell et al 1993) and *cdc2* (Ku et al 1993). Finally, Myb also plays a role in regulating haemopoietic cell differentiation (Weber et al 1990). It functions as a transcription factor for several cellular genes including those encoding the neutrophil granule protein mim-1 (Ness et al 1989), CD4 (Nakayama et al 1993), insulin-like growth factor 1 (IGF-1) (Travali et al 1991) and CD34 (Melotti et al 1994), and possibly other growth factors (Szczylik et al 1993) including c-*kit* (Ratajczak et al 1992c,d). The latter is of particular interest since it has been shown that when haemopoietic cells are deprived of c-*kit* ligand (Steel Factor) they undergo apoptosis (Yu et al 1993). Accordingly, c-*myb* is clearly an important haemopoietic cell gene which may, directly or indirectly, contribute to the pathogenesis or maintenance of human leukaemias. For this reason it is a rational target for therapeutically motivated disruption strategies.

Targeting c-*myb*

Our investigations were initially designed to elucidate the role of Myb in regulating haemopoietic cell development. Because the results obtained from these studies had obvious clinical relevance, more translationally oriented studies were also undertaken. These have now culminated in ongoing clinical trials at the Hospital of the University of Pennsylvania. Below I summarize the steps carried out in the clinical development of the c-*myb*-targeted antisense oligodeoxynucleotide. In addition, I will also allude briefly to our initial clinical experience with the *myb*-targeted oligodeoxynucleotide.

In vitro experience in the haemopoietic cell system

Role of c-myb-encoded protein in normal human haemopoiesis.   Attempts to exploit the c-*myb* gene as a therapeutic target for antisense oligodeoxynucleotides began as an outgrowth of studies that were seeking to define the role of Myb in regulating normal human haemopoiesis (Gewirtz et al 1989, Gewirtz & Calabretta 1988). During the course of these studies it was shown that exposing normal bone marrow mononuclear cells to c-*myb* antisense oligodeoxynucleotides resulted in a decreased cloning efficiency and progenitor cell proliferation. The effect was lineage indifferent, since c-*myb* antisense DNA inhibited granulocyte macrophage colony-forming units (CFU-GM), CFU-E (erythroid) and CFU-Meg (megakaryocyte). In contrast, a c-*myb* oligodeoxynucleotide with the corresponding sense sequence had no consistent effect on haemopoietic colony formation when compared to growth in control cultures. Finally, inhibition of colony formation was also dose dependent. Inhibition of the targeted mRNA was also demonstrated. Sequence-specific, dose-dependent biological effects accompanied by a specific decrease or total elimination of the targeted mRNA strongly suggested that the effects we were observing were due to an antisense mechanism. It should be added that the effects we observed were largely confirmed using homologous recombination (Mucenski et al 1991). In other investigations, it was also determined that haemopoietic progenitor cells require Myb during specific stages of development, in particular when they were actively cycling (Caracciolo et al 1990), as might be expected given the above functional description of Myb.

Myb is also required for leukaemic haemopoiesis.   Since the c-*myb* antisense oligodeoxynucleotide inhibited normal cell growth, we were also interested in determining its effect on leukaemic cell growth. While one could reasonably postulate that aberrant c-*myb* expression or Myb function might play a role in carcinogenesis, demonstrating this is another matter. To address this question, we employed a variety of leukaemic cell lines, including those of myeloid and lymphoid origin. In addition, we also employed primary patient material. We first determined the effect of c-*myb* sense and antisense oligodeoxynucleotides on the growth of HL-60,

K562, KG-1 and KG-1a myeloid cell lines (Anfossi et al 1989). The antisense oligodeoxynucleotide inhibited the proliferation of each leukaemic cell line, although the effect was most pronounced in HL-60 cells. Specificity of this inhibition was demonstrated by the observation that the sense oligodeoxynucleotide had no effect on cell proliferation, nor did antisense sequences with two or four nucleotide mismatches. To determine whether the treatment with c-*myb* antisense oligodeoxynucleotide modified the cell cycle distribution of HL-60 cells, we measured the DNA content in exponentially growing cells exposed to either sense or antisense c-*myb* oligodeoxynucleotides. Control cells, and those treated with c-*myb* sense oligodeoxynucleotide, had twice the DNA content of HL-60 cells exposed to the antisense oligodeoxynucleotide. The majority of these cells appeared to reside either in the G1 compartment or were blocked at the G1/S boundary. To examine the effect of the c-*myb* oligodeoxynucleotide on lymphoid cell growth, we employed a lymphoid leukaemia cell line, CCRF-CEM. As we noted in the case of normal lymphocytes (Gewirtz et al 1989), the CCRF-CEM cells were extremely sensitive to the anti-proliferative effects of the c-*myb* antisense oligodeoxynucleotide, whereas there were negligible effects of sense oligodeoxynucleotide on CEM cell growth in short-term suspension cultures. Exposure to c-*myb* antisense DNA resulted in a daily decline in cell numbers: compared to untreated controls, antisense DNA inhibited growth by ~2 log. Growth reduction was not a cytostatic effect, since cell viability was reduced by only ~70% after exposure to the antisense oligodeoxynucleotide, and CEM cell growth did not recover when cells were left in culture for an additional nine days.

Results obtained from primary patient material were equally encouraging (Table 1) (Calabretta et al 1991). We began by attempting to determine whether CFU-L (leukaemic) from AML patients could be inhibited by exposure to c-*myb* antisense oligodeoxynucleotide. Of the 28 patients we initially studied, colony and cluster data were available in 16 and 23 cases, respectively. After exposure to relatively low doses of c-*myb* antisense oligodeoxynucleotide (60 μg/ml) colony formation was inhibited in a statistically significant manner in 12/16 (~75%) cases. Cluster formation was similarly inhibited. The numbers of residual colonies in the antisense treated dishes was ~10%.

An obvious problem with interpreting these results, however, was determining the nature of the residual cells, i.e. were they the progeny of residual normal CFU or CFU-L? To try to answer this question in a rigorous manner, we turned out attention to chronic myelogenous leukaemia (CML) where the presence of the t(9:22) or Bcr–Abl neogene provided an unequivocal marker of the malignant cells (Ratajczak et al 1992a). Exposure of CML cells to c-*myb* antisense oligodeoxynucleotide resulted in the inhibition of CFU-GM-derived colony formation in >50% cases evaluated and thus far we have studied in excess of 40 patients. Representative data are shown in Fig. 2. These are presented as a function of oligomer effect on cells, with 'high' cloning efficiency (control colonies >250/plate) (Fig. 2A) versus 'low' cloning efficiency (control colonies <250/plate) (Fig. 2B). In this particular study, colony formation was observed in eight of 11 cases evaluated and was statistically significant ($P < 0.03$)

TABLE 1   Effect of c-myb oligomers on primary acute myelogenous leukaemia (AML) cell colony/cluster formation

Case	Colonies (% of control)		Clusters (% of control)	
	Sense	Antisense	Sense	Antisense
1	86	18 (0.058)	60	37 (0.080)
4	NG	NG	90	28 (0.036)
5	NG	NG	70	22 (0.101)
6	NG	NG	79	22 (0.026)
7	170	100 (0.423)	76	128 (0.502)
8	92	11 (0.008)	96	46 (0.020)
10	NG	NG	190	216 (0.034)
11	45	14 (0.021)	58	21 (0.084)
14	68	1 (0.152)	90	53 (0.071)
15	66	81 (0.736)	100	100 (0.896)
16	NG	NG	66	24 (0.001)
17	NG	NG	16	8 (0.023)
18	NG	NG	110	77 (0.164)
19	113	116 (0.717)	91	91 (0.763)
20	92	9 (0.051)	100	50 (0.009)
21	94	0 (0.006)	90	06 (0.004)
22	80	13 (0.001)	103	11 (0.015)
23	63	6 (0.001)	74	27 (0.004)
24	87	17 (0.002)	91	26 (0.018)
25	100	0 (0.019)	107	38 (0.364)
26	76	0 (0.009)	89	00 (0.001)
27	79	21 (0.014)	59	18 (0.043)
28	88	20 (0.009)	94	152 (0.096)

Blast cells were isolated from the peripheral blood of AML patients and exposed to sense or antisense oligomers. Colonies and clusters were enumerated and values were compared with growth in control cultures, which contained no oligomers. For each case, the number of colonies or clusters arising in the untreated control dishes was assumed to represent maximal (100%) growth for that patient. The numbers of colonies or clusters arising in the oligomer-treated dishes are expressed as a percentage of this number. NG, no growth. The statistical significance (determined by Student's t test for unpaired samples) of the change observed in the antisense-treated dishes is given as a P value in parentheses.

in seven. The amount of inhibition seen was dose dependent and ranged between 58% and 93%. In two cases the effect of the c-*myb* oligomers on CFU-GEMM (granulocyte, erythrocyte, monocyte, megakaryocyte) colony formation was also determined for the assessment of the effect of the oligomers on progenitors more primitive than

CFU-GM. In each case, significant inhibition of CFU-GEMM-derived colony formation was noted. It is also important to note that colony inhibition was sequence specific. For example, as shown in Fig. 2, c-*myb* sense oligodeoxynucleotide fails to inhibit colony formation significantly when employed at the highest antisense doses utilized.

Normal and leukaemic progenitor cells rely differentially on c-myb function.   In order to be useful as a therapeutic target, leukaemic cells would have to be more dependent on Myb than their normal counterparts. To examine this critical issue, we incubated phagocyte- and T cell-depleted normal human marrow mononuclear cells, human T lymphocyte leukaemia cell line blasts (CCRF-CEM) or 1:1 mixtures of these cells with sense or anti-sense oligodeoxynucleotides to codons 2–7 of human c-*myb* mRNA(Calabretta et al 1991). Oligodeoxynucleotides were added to liquid suspension cultures at Time 0 and after 18 h. Control cultures were untreated. In controls, or in cultures to which 'high' doses of sense oligodeoxynucleotides were added, CCRF-CEM proliferated rapidly, whereas mononuclear cells numbers and viability decreased by <10%. In contrast, when CCRF-CEM were incubated for four days in c-*myb* antisense DNA, cultures contained $4.7 \pm 0.8 \times 10^4$ cells/ml (mean $\pm$ S.D.; $n = 4$) compared to $285 \pm 17 \times 10^4$ cells/ml in controls. At the effective antisense dose, mononuclear cells were largely unaffected. After four days in culture, the remaining cells were transferred to methylcellulose supplemented with recombinant haemopoietic growth factors. Myeloid colonies/clusters were enumerated at Day 10 of culture inception. Depending on cell number plated, control mononuclear cells formed from 31 ± 4 to 274 ± 18 colonies. In dishes containing equivalent numbers of untreated or sense oligodeoxynucleotide-exposed CCRF-CEM, colonies were too numerous to count. When mononuclear cells were mixed 1:1 with CCRF-CEM in antisense oligomer concentrations $<5\,\mu g/ml$, only leukaemic colonies could be identified by morphological, histochemical and immunochemical analysis. However, when antisense oligomer exposure was intensified, normal myeloid colonies could now be found in the culture, whereas leukaemic colonies could no longer be identified with certainty using the same analytical methods. Finally, at antisense DNA doses used in the above studies, AML blasts from 18 out of 23 patients exhibited $\sim 75\%$ decrease in colony and cluster formation compared to untreated or sense oligomer-treated controls. When 1:1 mixing experiments were carried out with primary AML blasts and normal mononuclear cells, we were again able to preferentially eliminate AML blast colony formation, whereas normal myeloid colonies continued to form.

Use of c-myb oligodeoxynucleotides as bone marrow-purging agents.   The above experiments suggested that leukaemic cell growth could be preferentially inhibited after exposure to c-*myb* antisense oligodeoxynucleotides. In contemplating a clinical use for our findings, application in the area of bone marrow transplantation seemed compelling. In this application, exposure conditions are entirely under the control of the investigator. In addition, the patient's exposure to the antisense DNA is minimal.

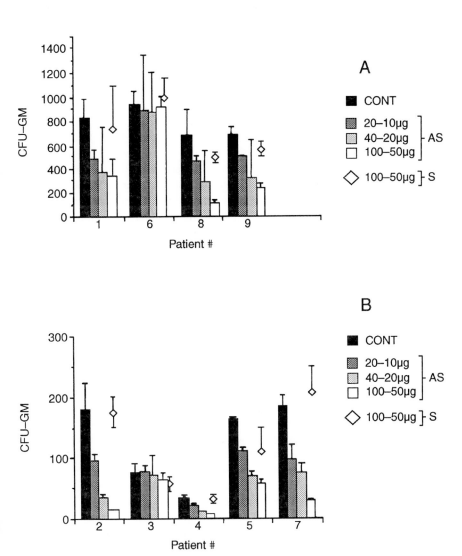

FIG. 2. Effect of c-*myb* oligomers on chronic myeloid leukaemia cell colony formation by cells with 'high' (A) and 'low' (B) cloning efficiency. Colony-forming cells were enriched from patient peripheral blood or bone marrow and exposed to oligomers as detailed in the text. At 24 h cells were plated and resulting colonies were enumerated in plates containing untreated control cells, antisense (AS)- and sense (S)-treated cells. Values plotted are mean ± S.D. of actual colony counts compared to growth in control cultures which contained no oligomers. CFU-GM, granulocyte macrophage colony-forming units.

This circumstance would also make approval by regulatory agencies less difficult. We therefore determined if the antisense oligodeoxynucleotides could be utilized as *ex vivo* bone marrow-purging agents.

To examine this issue, we mixed normal mononuclear cells (1 : 1) with primary AML or CML blast cells and exposed them to the oligodeoxynucleotides using a slightly modified protocol designed to test the feasibility of a more intensive antisense exposure. With this in mind, an additional oligodeoxynucleotide dose (20 μg/ml) was given just prior to plating the cells in methylcellulose. In control growth factor-stimulated cultures leukaemic cells formed 25.5 ± 3.5 (mean $\pm$ S.D.) colonies and 157 ± 8.5 clusters (per 2×10^5 cells plated). Exposure to c-*myb* sense oligodeoxynucleotides did not significantly alter these numbers (19.5 ± 0.7 colonies and 140.5 ± 7.8 clusters; $P > 0.1$). In contrast, equivalent concentrations of antisense oligodeoxynucleotides totally inhibited colony and cluster formation by the leukaemic blasts. Colony formation was also inhibited in the plates containing normal mononuclear cells, but only by $\sim 50\%$ in comparison to untreated control plates (control colony formation, 296 ± 40 per 2×10^5 cells plated; treated colony formation, 149 ± 15.5 per 2×10^5 cells).

To assess the potential effectiveness of an antisense purge, we carried out co-culture studies with cells obtained from CML patients in blast crisis and in chronic phase of their disease (Ratajczak et al 1992a). CML was a particularly useful model because cells from the malignant clone carry a tumour-specific chromosomal translocation that can be easily identified in tissue culture by looking for *bcr–abl*, the mRNA product of the gene produced by the translocation (Witte 1993). RNA was therefore extracted from cells cloned in methylcellulose cultures after exposure to the highest c-*myb* antisense oligodeoxynucleotide dose. The RNA was then reverse transcribed and the resulting cDNA amplified. For each patient studied, mRNA was also extracted from a comparable number of cells derived from untreated control colonies using the same technique. Eight cases were evaluated and in each case *bcr–abl* expression as detected by reverse transcriptase (RT)-PCR correlated with colony growth in cell culture. In cases that were inhibited by exposure to c-*myb* antisense oligodeoxynucleotides (7/11), *bcr–abl* expression was also greatly decreased or non-detectable (Fig. 3A). These results suggested that *bcr–abl* expressing CFU might be substantially or entirely eliminated from a population of blood or marrow mononuclear cells by exposure to the antisense oligodeoxynucleotides. To explore this possibility further, we carried out a re-plating experiment on samples from two patients (Fig. 3B). We hypothesized that if CFU belonging to the malignant clone were present at the end of the original 12-day culture period, but not detectable because of failure to express *bcr–abl*, they might re-express the message upon re-growth in fresh cultures. Accordingly, cells from these patients were exposed to oligodeoxynucleotides and then plated into methylcellulose cultures formulated to favour growth of either CFU-GM or CFU-GEMM. As was found with the original specimens, untreated control cells and cells exposed to sense oligodeoxynucleotides had detectable *bcr–abl* transcripts as determined by RT-PCR. Those exposed to the c-*myb* antisense oligodeoxynucleotides had none. One of the

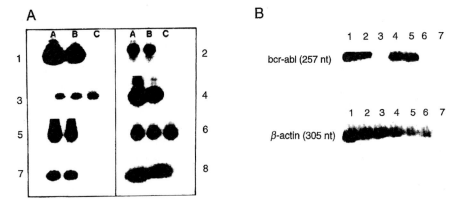

FIG. 3. (A) Detection of *bcr–abl* transcripts in colonies derived from granulocyte macrophage colony-forming units (CFU-GM) obtained from marrow of eight patients whose marrow was unexposed to A (c-*myb* antisense oligodeoxynucleotide), B (c-*myb* sense oligodeoxynucleotide) or C (c-*myb* antisense oligodeoxynucleotide). All colonies present in the variously treated methylcellulose cultures were harvested and subjected to analysis. Colony selection bias was therefore avoided. (B) Detection of *bcr–abl* transcripts in colonies derived from CFU-GM (lanes 1–3) and CFU-GEMM (granulocyte, erythrocyte, monocyte, megakaryocyte) (lanes 4–6) obtained from re-seeded primary colonies of Patient 8. Note that while β-actin transcripts are clearly detected in all colony samples, *bcr–abl* is only detectable in colonies derived from untreated control colonies (lanes 1 & 4) and colonies previously exposed to Myb sense oligodeoxynucleotides (lanes 2 & 5). Cells derived from colonies originally exposed to c-*myb* antisense oligodeoxynucleotides do not have detectable *bcr–abl*-expressing cells. Lane 7 is a control lane for the PCR reactions and is appropriately empty.

paired dishes from these cultures was solubilized with fresh medium, and all cells contained therein were washed, disaggregated and re-plated into fresh methylcellulose cultures without re-exposing the cells to oligodeoxynucleotides. After 14 days, CFU-GM and CFU-GEMM colony cells were again probed for *bcr–abl* expression. Control and sense-treated cells had detectable mRNA as determined by RT-PCR but none was found in the antisense-treated colonies. These results suggest that elimination of *bcr–abl*-expressing cells and CFU was highly efficient and perhaps permanent.

Efficacy of c-myb oligodeoxynucleotides in vivo: development of animal models

The studies described above were carried out primarily with unmodified DNA. Such molecules are subject to endonuclease and exonuclease attack at the phosphodiester bonds and are therefore of little utility *in vivo*. We therefore needed to address two questions at this point. First, we needed to know if a more stable, chemically modified oligodeoxynucleotide would give similar results. Second, we needed to know if these materials would be effective against human leukaemia cells in an *in vivo* system. Since we could not give this material to patients we established a human leukaemia/SCID mouse model system which would allow us to address both questions simultaneously (Ratajczak et al 1992b). To carry out these experiments, we intravenously injected SCID mice with K562 chronic myeloid leukaemia cells after cyclophosphamide conditioning. K562 cells express c-*myb*, the antisense oligodeoxynucleotide target and the tumour-specific Bcr–Abl oncogene, which was utilized for tracking the human leukaemia cells in the mouse host. After tumour cell injection, animals developed blasts in the peripheral blood within four to six weeks. After peripheral blood blast cells appeared, mean (S.D.) survival of untreated mice ($n = 20$) was 6 ± 3 days. Dying animals had prominent CNS infiltration, marked infiltration of the ovary and scattered abdominal granulocytic sarcomas. Infusion of either sense or scrambled sequence c-*myb* phosphorothioate oligodeoxynucleotides (24 bp; codons 2–9) for three, seven or 14 days had no statistically significant effect on sites of disease involvement, or on animal survival compared with control animals. In contrast, animals treated for seven or 14 days with c-*myb* antisense oligodeoxynucleotides survived 3.5- to eightfold longer ($P < 0.001$) than the various control animals ($n = 60$) (Fig. 4). In addition, animals receiving c-*myb* antisense DNA had either rare microscopic foci or no obviously detectable CNS disease (Fig. 3), and a $> 50\%$ reduction of ovarian involvement. A three-day infusion of c-*myb* antisense oligonucleotides (100 μg/day) was without effect. Infusing mice ($n = 12$) with antisense oligodeoxynucleotides (200 μg/day for 14 days) complementary to the c-Kit proto-oncogene, which K562 cells do not express, also had no effect on disease burden or survival ($n = 12$). These results suggested that phosphorothioate-modified c-*myb* antisense DNA might be efficacious for the treatment of human leukaemia *in vivo*.

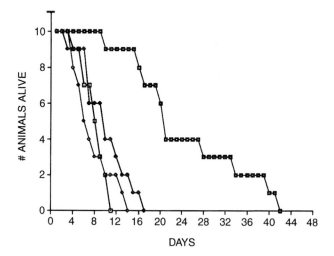

FIG. 4.   Survival curves of SCID–human chimeric animals transplanted with K562 chronic myelogenous leukaemia cells. Animals received a 14-day infusion of oligomers at a dose of 100 μg/day. □, antisense; ◆, sense; ⊡, control; ◇, scrambled.

Why does down-regulating myb kill leukaemic cells preferentially? A hypothesis

Our initial studies on the function of the c-*kit* receptor in haemopoietic cells suggested that c-*kit* might be a Myb-regulated gene (Ratajczak et al 1992c). Since c-*kit* encodes a critical haemopoietic cell tyrosine kinase receptor (Ratajczak et al 1992d), we have hypothesized that dysregulation of c-*kit* expression may be an important mechanism of action of *myb* antisense oligodeoxynucleotides. In support of this hypothesis it has been shown that when haemopoietic cells are deprived of c-*kit* R ligand (Steel Factor) they undergo apoptosis (Yu et al 1993). It has also recently been shown that when CD56[bright] natural killer (NK) cells, which express c-*kit*, are deprived of their ligand (Steel Factor), they too undergo apoptosis, perhaps because *bcl-2* is down-regulated (Carson et al 1994). Malignant myeloid haemopoietic cells, in particular CML cells, also express c-*kit* and respond to Steel Factor. Accordingly, we postulate that perturbation of *myb* expression in malignant haemopoietic cells may force them to enter an apoptotic pathway by down-regulating c-*kit*. Preliminary studies of K562 cells exposed to c-*myb* antisense oligodeoxynucleotides demonstrate that such cells do in fact undergo nuclear degenerative changes characteristic of apoptosis (Fig. 5).

Of necessity, we must also postulate that normal progenitor cells, at least at some level of development, are more tolerant of this transient disturbance. Since neither

FIG. 5.   c-*myb* antisense oligodeoxynucleotides cause K562 cells to undergo apoptosis. K562 cells were exposed to c-*myb* sense or antisense oligodeoxynucleotide ($\sim 20\mu$M) for 36 h. After this time DNA was extracted from the cells, subjected to electrophoresis and stained with ethidium bromide. Lane 1: untreated control; nuclear DNA is intact. Lane 2: cells exposed to c-*myb* antisense oligodeoxynucleotides; note characteristic 'laddering' of DNA. Lane 3: cells exposed to c-*myb* sense oligodeoxynucleotides.

Steel nor White Spotting (*W*) mice (which lack the *kit* receptor or its ligand, respectively) are aplastic, this is a tenable hypothesis (Ratajczak et al 1992d).

Pharmacodyamic studies with a *myb*-targeted oligodeoxynucleotides

As we have noted in the past, understanding uptake mechanisms and intracellular handling might allow oligodeoxynucleotides to be used with enhanced biological effectiveness. We have examined these issues at the ultrastructural level in the hope of gaining information that will be useful for oligonucleotide design and rational administration in the clinic (Beltinger et al 1995).

We first sought to identify oligodeoxynucleotide-binding proteins on haemopoietic cell surfaces (Fig. 6). We then attempted to follow oligodeoxynucleotide trafficking once inside the cell. To conduct these studies, K562 cells were incubated at 4 °C with biotin-labelled oligodeoxynucleotides alone or excess unlabelled oligodeoxynucleotides of identical sequence. The identification of binding proteins involved cross-linking them to bound oligodeoxynucleotides, extracting the complexes with 2% Nonidet P-40 and resolving them on SDS-PAGE gels. This process is then followed by colorometric detection. In contrast to previous reports (Loke et al 1989), we identified at least five major oligodeoxynucleotide-binding proteins ranging in size from $\sim 20\,kDa$ to $143\,kDa$. Excess unlabelled sense oligodeoxynucleotides (500-fold), but not free biotin, inhibited phosphorothioate oligodeoxynucleotide binding, demonstrating the specificity of binding proteins. Neuraminidase treatment of the cells prior to incubation with oligodeoxynucleotides decreased protein binding to most oligodeoxynucleotides, suggesting that sugar moieties play a role in binding. In other experiments, oligodeoxynucleotide-binding proteins were examined after metabolic labelling with [35S]methionine. Binding proteins of similar number, migration pattern and relative abundance were again identified after gel resolution under non-denaturing conditions. Thus, the oligodeoxynucleotide-binding proteins on haemopoietic cells are not composed of subunits.

To visualize receptor-mediated endocytosis, we incubated K562 cells at 4 °C with biotinylated oligodeoxynucleotides. After cross-linking, cells were incubated with gold–streptavidin particles, warmed to 37 °C, fixed, and then processed for electron microscopy. Oligodeoxynucleotides were clearly identified in clathrin-coated pits, consistent with receptor-mediated endocytosis. We also carried out electron microscopic studies employing gold–streptavidin particles and [Sb]antisense oligodeoxynucleotides to visualize intracellular trafficking. Oligodeoxynucleotides were seen in endosomes, lysosome-like bodies and throughout the cytoplasm. A significant amount of labelled material was also observed in the nucleus.

Finally, in order to document an 'antisense' mechanism, it is necessary to correlate mRNA levels of the targeted gene with biological effects observed. We have demonstrated that this is feasible and that one can adequately follow target gene

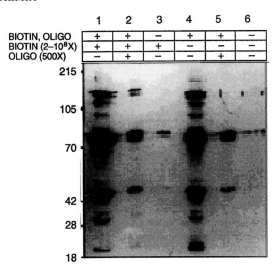

	1	2	3	4	5	6
BIOTIN, OLIGO	+	+	−	+	+	−
BIOTIN (2–10⁸X)	+	+	+	−	−	−
OLIGO (500X)	−	+	−	−	+	−

FIG. 6. K562 cells were incubated at 4 °C with biotin-labelled phosphorothioate oligodeoxynucleotides. Proteins to which the phosphorothioate oligodeoxynucleotides bound were identified by chemical cross-linking with BS3, followed by extraction of the complexes with 2% Nonidet P-40. These were subsequently resolved on SDS-PAGE gels and identified colorometrically by Western blotting using chemiluminescence. We identified five doublet bands representing oligodeoxynucleotide-binding proteins ranging in size from ~20 kDa to ~143 kDa (lanes 1 and 4). Excess unlabelled phosphorothioate oligodeoxynucleotides (lanes 2 and 5), but not free biotin (lane 1), inhibited phosphorothioate oligodeoxynucleotide binding, suggesting specificity of binding protein interactions. D-biotin not conjugated to oligodeoxynucleotides could not be detected bound to oligodeoxynucleotide-binding protein (lane 2). Molecular weight markers are depicted on the left. Lane 6 is empty.

mRNA levels in tissues of animals receiving oligodeoxynucleotides (Fig. 7) (Hijiya et al 1994).

Use of antisense oligonucleotides in a clinical setting

For the purpose of developing an antisense oligonucleotide therapeutic, CML seemed to us to be an excellent disease model. As mentioned above, CML is relatively common and it has a convenient marker chromosome and gene for objectively following potential therapeutic efficacy of a test compound (Gale et al 1993). In addition to these considerations, CML is uniformly fatal, except for individuals who are fortunate enough to have an allogeneic bone marrow donor. Picking a gene target in CML was actually somewhat problematic. An obvious target was the *bcr–abl* mRNA (Fig. 8) (Melo 1996). However, because *bcr–abl* is not expressed in primitive

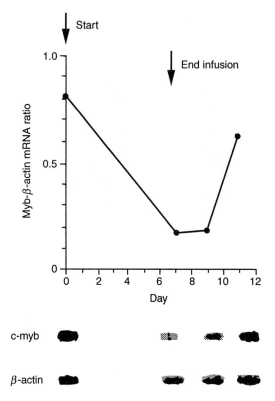

FIG. 7.   Effect of c-*myb* antisense oligodeoxynucleotides on c-*myb* mRNA expression in tumour tissue obtained from human melanoma-bearing SCID mice. Mice were infused with the antisense oligodeoxynucleotides (500 μg/day for seven days) and tumours were excised on Days 7, 9 and 11 after infusions were begun. c-*myb* and β-actin mRNA was detected in the same tumour tissue sample (∼1 g) by semi-quantitative reverse transcriptase PCR and quantitated by scanning densitometry. The relative amount of c-*myb* mRNA in each sample was estimated by normalization to the actin mRNA present in each sample.

haemopoietic stem cells (Bedi et al 1993), and because it is uncertain that transient interruption of Bcr–Abl signalling actually results in the death of CML cells, we felt that an alternative target might be of greater use in treating this disease. Based on the type of data presented above, a favourable therapeutic index in toxicology testing and more detailed knowledge of the pharmacokinetics of oligonucleotides, we have begun to evaluate the c-*myb*-targeted antisense oligodeoxynucleotide in the clinic (Gewirtz et al 1996a).

Towards this end, we initiated clinical trials to evaluate the effectiveness of phosphorothioate-modified antisense oligodeoxynucleotides to c-*myb* as marrow-

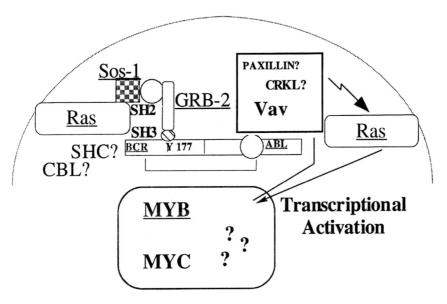

FIG. 8. Molecular pathogenesis of chronic myelogenous leukaemia. Bcr–Abl, product of the gene *bcr–abl* created by the t(9:22) co-operates with a number of signalling partners to activate Ras. Ras activation is postulated to play a critical role in haemopoietic cell transformation.

purging agents for chronic-phase or accelerated-phase CML patients, and have undertaken a Phase I intravenous infusion study for blast crisis patients and patients with other refractory leukaemias. Oligodeoxynucleotide purging was carried out for 24 h on CD34$^+$ marrow cells. Patients received busulfan and cytoxan, followed by re-infusion of previously cryopreserved phosphorothioate oligodeoxynucleotide-purged mononuclear cells. In the pilot marrow-purging study seven chronic phase and one acute phase CML patients have been treated. Engrafting occurred in seven of these patients. In four out of six evaluable chronic-phase patients, metaphases were 85–100% normal three months after engraftment, suggesting that a significant purge had taken place in the marrow graft. Five chronic-phase patients have demonstrated marked, sustained, haematological improvement with essential normalization of their blood counts. The follow-up period ranged from six months to two years. In an attempt to further increase purging efficiency we incubated patient mononuclear cells for 72 h in the P-oligodeoxynucleotide. Although PCR and long-term culture-initiating cell studies suggested that an efficient purge had occurred, engraftment in five patients was poor. In the Phase I systemic infusion study of 18 refractory leukaemia patients (two patients were treated at two different dose levels; 13 had acute-phase or blast crisis CML) *myb* antisense oligodeoxynucleotides were delivered by continuous infusion at dose levels ranging between 0.3 mg/kg/day for seven days to 2.0 mg/kg/day for seven days. No recurrent dose-related toxicity has been noted, although idiosyncratic toxicities that are clearly not drug related were observed (one transient renal insufficiency; one pericarditis). One blast crisis patient survived for about 14 months with transient restoration of chronic-phase disease. These studies show that oligodeoxynucleotides may be administered safely to patients with leukaemia. Whether patients treated on either study derived clinical benefit is uncertain, but the results of these studies suggest that oligodeoxynucleotides may eventually demonstrate therapeutic utility in the treatment of human leukaemias.

Problem solving

The power of the antisense approach has been demonstrated in experiments in which critical biological information has been gathered using antisense technology and has been subsequently verified by other laboratories using other methodologies (Gewirtz & Calabretta 1988, Metcalf 1994, Mucenski et al 1991). However, this technology, in spite of its successes, has been found to be highly variable in its efficiency. To the extent that many have tried to employ oligodeoxynucleotides and more than a few have been perplexed and frustrated by results that were non-informative at best, or even worse, misleading or unreproducible, it is easy to understand why this approach has become somewhat controversial. We believe that progress on two fronts would help address this problem.

First, in order for an oligodeoxynucleotide to hybridize with its mRNA target, it must find an accessible sequence. Sequence accessibility is at least in part a function of mRNA physical structure, which is dictated in turn by internal base composition and associated proteins in the living cell. Attempts to describe the *in vivo* structure of RNA, in contrast to DNA, have been fraught with difficulty (Baskerville & Ellington 1995). Accordingly, mRNA targeting is largely a hit or miss process, accounting for many experiments where the addition of an oligodeoxynucleotide yields no effect on expression. Hence, the ability to determine which regions of a given mRNA molecule are accessible for oligodeoxynucleotide targeting is a significant impediment to the application of this technique in many cell systems. We have begun to approach this issue by developing a footprinting assay to determine which physical areas of an RNA are accessible to the oligonucleotide. We have proceeded under the assumption that sequences which remain accessible to single-stranded RNases in a more physiological environment may also remain accessible for hybridization to an oligodeoxynucleotide. Preliminary experiments performed in our laboratory in which a labelled RNA transcript is allowed to hybridize with an oligonucleotide in the presence or absence of nuclear extracts from the cells of interest along with RNase T1 suggest that footprinting of this type is feasible (Fig. 9). Of more interest, our preliminary results suggest that this approach may be of use in designing oligonucleotides.

Second, the ability to deliver oligodeoxynucleotides into cells and have them reach their target in a bioavailable form also remains problematic (Gewirtz et al 1996b). Without this ability, it is clear that even an appropriately targeted sequence is not

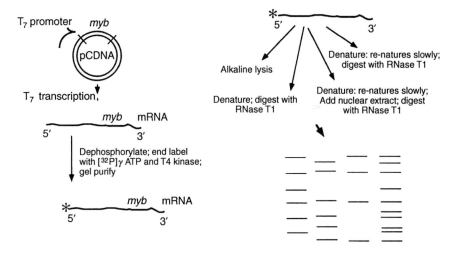

FIG. 9.   Schematic diagram of RNA footprinting methodology for designing oligonucleotides.

likely to be efficient. Native phosphodiester oligodeoxynucleotides, and the widely used phosphorothioate-modified oligodeoxynucleotides, which contain a single sulfur substituting for oxygen at a non-bridging position at each phosphorus atom, are polyanions. Accordingly, they diffuse across cell membranes poorly and are only taken up by cells through energy-dependent mechanisms. This appears to be accomplished primarily through a combination of adsorbtive endocytosis and fluid phase endocytosis which may be triggered in part by the binding of the oligodeoxynucleotide to receptor-like proteins present on the surface of a wide variety of cells (Beltinger et al 1995, Loke et al 1989). Confocal and electron microscopy studies have indicated that after internalization, the bulk of the oligodeoxynucleotides enter the endosome/lysosome compartment. These vesicular structures may become acidified and acquire other enzymes that degrade the oligodeoxynucleotides. Biological inactivity is the predictable result of this process. Recently described strategies for introducing oligodeoxynucleotides into cells, including various cationic lipid formulations, may address this problem (Fig. 10) (Bergan et al 1996, Lewis et al 1996, Spiller & Tidd 1995).

Conclusions

The ability to block gene function with antisense oligodeoxynucleotides has become an important tool in many research laboratories. Since activation and aberrant expression of proto-oncogenes appears to be an important mechanism in malignant transformation, targeted disruption of these genes and other molecular targets with oligodeoxynucleotides could have significant therapeutic utility as well. In this

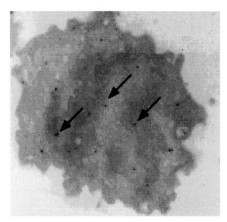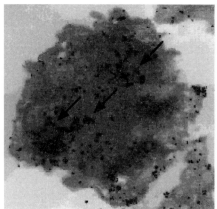

FIG. 10. Uptake of phosphorothioate oligodeoxynucleotides delivered by streptolysin permeabilization (left) or cationic lipid (right). Oligonucleotides were synthesized with biotin tags so that they could be decorated with gold–streptavidin particles which appear as black dots in the photomicrographs. Most of the material appears to be localized to the nucleus of the cells.

regard, the potential therapeutic usefulness of oligodeoxynucleotides has been demonstrated in many systems and against a number of different targets including viruses, oncogenes, proto-oncogenes and an increasing array of cellular genes. These studies suggest that synthetic oligodeoxynucleotides have the potential to become important new therapeutic agents for the treatment of human cancer. Nevertheless, it is clear that considerable optimization will be required before antisense oligonucleotides emerge as effective agents for treating human disease. Progress will need to occur on several fronts. These include issues related to the chemistry of the molecules employed; for example, finding out how chemical modifications impact on uptake, stability and hybridization efficiency of the synthetic DNA molecule. A clearer understanding of the mechanism of antisense-mediated inhibition, including where such inhibition takes place, will also be required. Finally, cellular defence mechanisms, such as increasing transcription of the targeted message, may also be factors to consider in planning effective treatment strategies with these agents. The choice of target is also an important issue. Although many issues remain to be resolved, we remain optimistic that this approach will one day prove useful for the treatment of patients with a variety of haematological malignancies.

Acknowledgements

This work was supported by grants from the National Institutes of Health and the Leukemia Society of America.

References

Anfossi G, Gewirtz, AM, Calabretta B 1989 An oligomer complementary to c-*myb*-encoded mRNA inhibits proliferation of human myeloid leukemia cell lines. Proc Natl Acad Sci USA 86:3379–3383

Barletta C, Pelicci PG, Kenyon LC, Smith SD, Dalla-Favera R 1987 Relationship between the c-*myb* locus and the 6q-chromosomal aberration in leukemias and lymphomas. Science 235:1064–1067

Baskerville S, Ellington AD 1995 RNA structure. Describing the elephant. Curr Biol 5:120–123

Bedi A, Zehnbauer BA, Collector MI et al 1993 Bcr–Abl gene rearrangement and expression of primitive hematopoietic progenitors in chronic myeloid leukemia. Blood 81:2898–2902

Beltinger C, Saragovi HU, Smith RM et al 1995 Binding, uptake, and intracellular trafficking of phosphorothioate-modified oligodeoxynucleotides. J Clin Invest 95:1814–1823

Bergan R, Hakim F, Schwartz GN et al 1996 Electroporation of synthetic oligodeoxynucleotides: a novel technique for *ex vivo* bone marrow purging. Blood 88:731–741

Biedenkapp H, Borgmeyer U, Sippel AE, Klempnauer KH 1988 Viral *myb* oncogene encodes a sequence-specific DNA-binding activity. Nature 335:835–837

Burk O, Mink S, Ringwald M, Klempnauer KH 1993 Synergistic activation of the chicken MIM-1 gene by v-Myb and C/EBP transcription factors. EMBO J 12:2027–2038

Calabretta B, Sims RB, Valtieri M et al 1991 Normal and leukemic hematopoietic cells manifest differential sensitivity to inhibitory effects of c-*myb* antisense oligodeoxynucleotides: an *in vitro* study relevant to bone marrow purging. Proc Natl Acad Sci USA 88:2351–2355

Caracciolo D, Venturelli D, Valtieri M, Peschle C, Gewirtz AM, Calabretta B 1990 Stage-related proliferative activity determines c-*myb* functional requirements during normal human hematopoiesis. J Clin Invest 85:55–61

Carson WE, Haldar S, Baiocchi RA, Croce CM, Caligiuri MA 1994 The c-*kit* ligand suppresses apoptosis of human natural killer cells through the upregulation of *bcl-2*. Proc Natl Acad Sci USA 91:7553–7557

Clarke MF, Kukowska-Latallo JF, Westin E, Smith M, Prochownik EV 1988 Constitutive expression of a c-*myb* cDNA blocks Friend murine erythroleukemia cell differentiation. Mol Cell Biol 8:884–892

Cogswell JP, Cogswell PC, Kuehl WM et al 1993 Mechanism of c-*myc* regulation by c-*myb* in different cell lineages. Mol Cell Biol 13:2858–2869

Dini PW, Eltman JT, Lipsick JS 1995 Mutations in the DNA-binding and transcriptional activation domains of v-*myb* cooperate in transformation. J Virol 69:2515–2524

Gale RP, Grosveld G, Canaani E, Goldman JM 1993 Chronic myelogenous leukemia: biology and therapy. Leukemia 7:653–658

Galli-Taliadoros LA, Sedgwick JD, Wood SA, Korner H 1995 Gene knock-out technology: a methodological overview for the interested novice. J Immunol Methods 181:1–15

Gewirtz AM, Calabretta B 1988 A c-*myb* antisense oligodeoxynucleotide inhibits normal human hematopoiesis *in vitro*. Science 242:1303–1306

Gewirtz AM, Anfossi G, Venturelli D, Valpreda S, Sims R, Calabretta B 1989 G1/S transition in normal human T-lymphocytes requires the nuclear protein encoded by *c-myb*. Science 245:180–183

Gewirtz AM, Luger S, Sokol D et al 1996a Oligodeoxynucleotide therapeutics for human myelogenous leukemia: interim results. Blood (suppl) 88:1069

Gewirtz AM, Stein CA, Glazer PM 1996b Facilitating oligonucleotide delivery: helping antisense deliver on its promise. Proc Natl Acad Sci USA 93:3161–3163

Heyer WD, Kohli J 1994 Homologous recombination. Experientia 50:189–191

Hijiya N, Zhang J, Ratajczak MZ et al 1994. Biologic and therapeutic significance of MYB expression in human melanoma. Proc Natl Acad Sci USA 91:4499–4503

Kanei-Ishii C, MacMillan EM, Nomura T et al 1992 Transactivation and transformation by Myb are negatively regulated by a leucine-zipper structure. Proc Natl Acad Sci USA 89: 3088–3092

Ku DH, Wen SC, Engelhard A et al 1993 c-myb transactivates cdc2 expression via Myb-binding sites in the 5′-flanking region of the human cdc2 gene (published erratum appears in J Biol Chem 1993 268:13010). J Biol Chem 268:2255–2259

Lewis JG, Lin K-Y, Kothavale A et al 1996 A serum-resistant cytofectin for cellular delivery of antisense oligodeoxynucleotides and plasmid DNA. Proc Natl Acad Sci USA 93:3176–3181

Liebermann DA, Hoffman-Liebermann B 1989 Proto-oncogene expression and dissection of the myeloid growth to differentiation developmental cascade. Oncogene 4:583–592

Loke SL, Stein CA, Zhang XH et al 1989 Characterization of oligonucleotide transport into living cells. Proc Natl Acad Sci USA 86:3474–3478

Luger SM, Ratajczak J, Ratajczak MZ et al 1996 A functional analysis of protooncogene Vav's role in adult human hematopoiesis. Blood 87:1326–1334

Lyon J, Robinson C, Watson R 1994 The role of Myb proteins in normal and neoplastic cell proliferation. Crit Rev Oncog 5:373–388

Melo JV 1996 The molecular biology of chronic myeloid leukaemia. Leukemia 10:751–756

Melotti P, Ku DH, Calabretta B 1994 Regulation of the expression of the hematopoietic stem cell antigen CD34: role of c-myb. J Exp Med 179:1023–1028

Metcalf D 1994 Blood. Thrombopoietin—at last. Nature 369:519–520

Morrow B, Kucherlapati R 1993 Gene targeting in mammalian cells by homologous recombination. Curr Opin Biotechnol 4:577–582

Mucenski ML, McLain K, Kier AB et al 1991 A functional c-Myb gene is required for normal murine fetal hepatic hematopoiesis. Cell 65:677–689

Nakayama K, Yamamoto R, Ishii S, Nakauchi H 1993 Binding of c-Myb to the core sequence of the CD4 promoter. Int Immunol 5:817–824

Ness SA, Marknell A, Graf T 1989 The v-Myb oncogene product binds to and activates the promyelocyte-specific Mim-1 gene. Cell 59:1115–1125

Ness SA, Kowenz-Leutz E, Casini T, Graf T, Leutz A 1993 Myb and NF-M: combinatorial activators of myeloid genes in heterologous cell types. Genes Dev 7:749–759

Nomura N, Zu YL, Maekawa T, Tabata S, Akiyama T, Ishii S 1993 Isolation and characterization of a novel member of the gene family encoding the cAMP response element-binding protein CRE-BP1. J Biol Chem 268:4259–4266

Nomura T, Sakai N, Sarai A et al 1993 Negative autoregulation of c-Myb activity by homodimer formation through the leucine zipper. J Biol Chem 268:21914–21923

Osman F, Tomsett B, Strike P 1994 Homologous recombination. Prog Ind Microbiol 29:687–732

Press RD, Reddy EP, Ewert DL 1994 Overexpression of C-terminally but not N-terminally truncated Myb induces fibrosarcomas: a novel nonhematopoietic target cell for the Myb oncogene. Mol Cell Biol 14:2278–2290

Prochownik EV, Smith MJ, Snyder K, Emeagwali D 1990 Amplified expression of three jun family members inhibits erythroleukemia differentiation. Blood 76:1830–1837

Ratajczak MZ, Hijiya N, Catani L et al 1992a Acute- and chronic-phase chronic myelogenous leukemia colony-forming units are highly sensitive to the growth inhibitory effects of c-myb antisense oligodeoxynucleotides. Blood 79:1956–1961

Ratajczak MZ, Kant JA, Luger SM et al 1992b In vivo treatment of human leukemia in a scid mouse model with c-myb antisense oligodeoxynucleotides. Proc Natl Acad Sci USA 89:11823–11827

Ratajczak MZ, Luger SM, DeRiel K et al 1992c Role of the KIT protooncogene in normal and malignant human hematopoiesis. Proc Natl Acad Sci USA 89:1710–1714

Ratajczak MZ, Luger SM, Gewirtz AM 1992d The c-Kit proto-oncogene in normal and malignant human hematopoiesis. Int J Cell Cloning 10:205–214

Sakura H, Kanei-Ishii C, Nagase T, Nakagoshi H, Gonda TJ Ishii S 1989 Delineation of three functional domains of the transcriptional activator encoded by the c-*myb* protooncogene. Proc Natl Acad Sci USA 86:5758–5762

Small D, Levenstein M, Kim E et al 1994 STK-1, the human homolog of *Flk-2/Flt-3*, is selectively expressed in CD34⁺ human bone marrow cells and is involved in the proliferation of early progenitor/stem cells. Proc Natl Acad Sci USA 91:459–463

Spiller DG, Tidd DM 1995 Nuclear delivery of antisense oligodeoxynucleotides through reversible permeabilization of human leukemia cells with streptolysin O. Antisense Res Dev 5:13–21

Szczylik C, Skorski T, Ku DH et al 1993 Regulation of proliferation and cytokine expression of bone marrow fibroblasts: role of c-*myb*. J Exp Med 178:997–1005

Takeshita K, Bollekens JA, Hijiya N, Ratajczak M, Ruddle FH, Gewirtz AM 1993 A homeobox gene of the *Antennapedia* class is required for human adult erythropoiesis. Proc Natl Acad Sci USA 90:3535–3538

Travali S, Reiss K, Ferber A et al 1991 Constitutively expressed c-*myb* abrogates the requirement for insulin-like growth factor 1 in 3T3 fibroblasts. Mol Cell Biol 11:731–736

Vorbrueggen G, Kalkbrenner F, Guehmann S, Moelling K 1994 The carboxy terminus of human c-*myb* protein stimulates activated transcription in trans. Nucleic Acids Res 22:2466–2475

Weber BL, Westin EH, Clarke MF 1990 Differentiation of mouse erythroleukemia cells enhanced by alternatively spliced c-*myb* mRNA. Science 249:1291–1293

Willnow TE, Herz J 1994 Homologous recombination for gene replacement in mouse cell lines. Methods Cell Biol 43:305–334

Witte ON 1993 Role of the Bcr–Abl oncogene in human leukemia: fifteenth Richard and Hinda Rosenthal Foundation Award Lecture. Cancer Res 53:485–489

Yu H, Bauer B, Lipke GK, Phillips RL, Van Zant G 1993. Apoptosis and hematopoiesis in murine fetal liver. Blood 81:373–384

DISCUSSION

Rossi: In the purging approach did you look at the fraction of the cell population by PCR analysis to see if you could purge Bcr–Abl before reinfusing the marrow cells into the patient?

Gewirtz: I tried to allude to that. If we do the PCR analysis on the post-purge cell fraction before it goes into the patient, in only one of those cases did we observe complete extinction of Bcr–Abl. The more I thought about this the more I thought that it was a remarkable result because when we get cells back from the blood bank they are a mixture of live and dead cells, and there's no doubt that cells expressing *bcr–abl* are present at various stages of life and/or death. Of more interest is the c-*myb* expression, which has an all-or-nothing expression pattern that doesn't necessarily relate to the clinical outcome. It is possible that we haven't studied enough patients, but my own surmise is that we are not looking at the population that counts, which is the stem cell population. This is a numbers problem because we're doing these experiments on a selected cell population, and by the time we do the harvesting we only have enough cells to do the transplant. Therefore, for ethical reasons it's a concern to take an aliquot of the treated cells for these experiments. My own belief is that the situation three months after the transplant should reflect the character of the cells that the patients were given.

Rossi: How does this compare with going through the regimen without the oligonucleotide?

Gewirtz: That's a fair question. One might expect to see cell reduction of this degree 20–25% of the time, and they are not long lived. We have only done this in a small number of patients, which is a caveat in this kind of a study.

Akhtar: Do you find that the patients who are also on drugs that compete for protein-binding sites on serum albumin, for example, have a better distribution of oligonucleotides to the target tissue?

Gewirtz: It's almost impossible to address issues like this in these patients because they are so sick, which means that they're receiving lots of different drugs all at once, and there are only three patients in each group.

Eckstein: I have two questions relating to the experiment involving RNase T1 digestion of the ^{32}P-labelled transcript in nuclear extracts. First, do RNases present in the nuclear extract interfere with the result? Second, why didn't you use a cellular extract? Because even though the oligonucleotides enter the nucleus that doesn't necessarily relate to activity.

Gewirtz: I did not present my results in chronological order, so we were still labouring under the misapprehension that the oligonucleotides had to be in the nucleus. Therefore, it would be perfectly reasonable to do this on a nucleus-free cell extract, and that's certainly something we would like to do. In answer to your first question, we add inhibitors to keep the RNA as stable as possible.

Southern: There didn't seem to be a large difference in footprint between tests with and without extracts.

Gewirtz: That's correct, although we believe that the result is significant because the proteins that the RNA associates with probably affect the folding of the RNA, which in turn affects its accessibility. We are just starting to develop a new technique, which may help us to answer this question.

Wickstrom: Do any of these cells typically overexpress the IGF-1 receptor?

Gewirtz: Haemopoetic cells certainly express reasonable amounts of IGF-1 receptor, but I'm not sure whether it's overexpressed or not. In any case we've shown that it is not required for growth.

Hélène: Ed Southern showed us yesterday that shifting the oligonucleotide by one nucleotide can completely change the hybridization (Southern et al 1997, this volume), so it is necessary to have a guanosine that is both accessible to RNase H and in the correct position. Therefore if the S2 and S3 oligonucleotides in your footprinting experiments were shifted by two or three nucleotides they might become active. What is the evidence that antisense oligonucleotides are in the cytoplasm or in the nucleus?

Gewirtz: That's a fair question. When I started out in this field five or six years ago the general feeling was that we were disrupting translation in the cytoplasm. However, data then accumulated suggesting that free oligonucleotides enter the nucleus. At least using the cell fraction subcellular fractionation studies, with all the problems that they have, we find little material in the cytoplasm and it seems to enter the nucleus. Therefore, the speculation was that they were really acting in the nucleus. It may be the case that we don't want them to act in the nucleus because there they may interact with too many proteins, and in which case it may be preferable to get them back into the cytoplasm, where they might be able to interact with the RNA.

Hélène: But has anyone shown that the antisense oligonucleotides are actually working in the nucleus?

Monia: We've been able to block splicing with an antisense oligonucleotide that acts through a non-RNase H mechanism. Since splicing occurs in the nucleus, the oligonucleotide must be acting in the nucleus.

Hélène: We have demonstrated that oligonucleotides work in the cytoplasm.

Monia: But we also know that oligonucleotides can work by arresting translation. This is not to say that they hybridize in the cytoplasm, rather they hybridize in the nucleus, exit it and then block translation in the cytoplasm. They could, however, work in both the cytoplasm and the nucleus.

Wagner: We have published the results of a pre-form assay that I described previously, where we inject oligonucleotides preformed onto RNA directly into the nucleus or the cytoplasm (Moulds et al 1995). We see 100% inhibition if we inject them in the former and 0% in the latter.

Matteucci: Was the sequence in your clinical trials the same anti-*myb* sequence that Rosenberg used in his anti-restenosis study (Simons et al 1992)?

Gewirtz: I don't believe so, but I think both sequences have a G_4 sequence.

Matteucci: He had an amazing effect but the antisense conclusion didn't stand up to other controls done subsequently by others (Burgess et al 1995).

Gewirtz: He stands by his results and thinks that experiments that were done to verify his results weren't done properly. He says that his results are reproducible.

Matteucci: Have you tried using his sequence, which clearly has a biological effect, in your biological assays?

Gewirtz: We have targeted downstream of the *myb* sequence but have not had any success. We would have to walk systemically along the RNA to see what works and what doesn't work.

Matteucci: You showed that with the lipid treatments you were getting oligonucleotide delivery but the phenotype was not enhanced. You interpreted this by saying that it was compartmentalized in the wrong compartment. Another interpretation is that your original effect was an interaction with a receptor at the surface of the cell.

Gewirtz: That is possible but we have looked at this extremely carefully. We are confident that we are getting an antisense effect with that oligonucleotide because we can down-regulate the mRNA and the protein.

Cohen: I just wanted to make the comment that, in terms of looking at subcellular distributions, I personally don't believe the results of microinjections. In my opinion, the results of microinjections are artefactual because of the streaming effect of non-protein-bound molecules. In order to show that nuclear uptake is occurring you really need to do confocal microscopy.

Gewirtz: I don't agree with this.

Cohen: The nuclear membrane is not the same as the cellular membrane, and streaming of molecules back and forth through the nuclear membrane probably occurs. Therefore, in terms of oligonucleotides working both in the cytoplasm and the nucleus, the nuclear membrane should not represent a significant barrier.

Rossi: Jack Cohen's point is a good one, i.e. that when you inject an oligonucleotide it probably has a different intracellular route than when it's taken up through some sort of an exogenous delivery system. Has anybody shown that injected oligonucleotides are functionally active?

Wagner: We have done these experiments, and we found that injected oligonucleotides are as active as oligonucleotides delivered by electroporation or cationic lipids.

Rossi: Did you show that the non-injected oligonucleotides entered the nucleus?

Wagner: In direct contrast to what Alan Gewirtz showed, we did not observe this.

Southern: We should think about what is happening within the nucleus because the nucleus is not a homogeneous compartment. I would also like to address the issue of whether we should be targeting the mRNA or the heterogeneous nuclear (hn) RNA. Has much work been done on hnRNA?

Lebleu: There has been little work on this, and what has been done has not fully taken into account what's known about splicing mechanisms.

References

Burgess TL, Fisher EF, Ross SL et al 1995 The antiproliferative activity of c-*myb* and c-*myc* antisense oligonucleotides in smooth muscle cells is caused by a nonantisense mechanism. Proc Natl Acad Sci USA 92:4051–4055

Moulds C, Lewis JG, Foehler BC et al 1995 Site and mechanism of antisense inhibition by C-5 propyne oligonucleotides. Biochemistry 34:5044–5053

Simons M, Edelman ER, De Keyser JL, Langer R, Rosenberg RD 1992 Antisense c-*myb* oligonucleotides inhibit intimal arterial smooth muscle cell accumulation *in vivo*. Nature 359:67–70

Southern EM, Milner N, Mir KU 1997 Discovering antisense reagents by hybridization of RNA to oligonucleotide arrays. In: Oligonucleotides as therapeutic agents. Wiley, Chichester (Ciba Found Symp 209) p 38–46

Therapeutic applications of catalytic antisense RNAs (ribozymes)

John J. Rossi

Department of Molecular Biology, Beckman Research Institute of the City of Hope, 1450 E Duarte Road, Duarte, CA 91010-3011, USA

Abstract. Ribozymes have progressed from an intriguing subject of scientific study to therapeutic agents for the potential treatment of a fatal and devastating viral infection. Despite this rapid road to clinical trials, there are many unexplored avenues that should be examined to improve the intracellular effectiveness of ribozymes. Since ribozymes are RNA molecules, the cellular rules governing RNA partitioning and stability can be applied to these molecules to make them more effective therapeutic agents. Future, successful therapeutic ribozyme applications will depend upon increasing our knowledge of RNA metabolism and movement, and applying this knowledge to the design of ribozymes. This chapter discusses experimental approaches towards this goal as well as recent progress in the application of a pair of hammerhead ribozymes for the clinical treatment of HIV1 infection.

1997 Oligonucleotides as therapeutic agents. Wiley, Chichester (Ciba Foundation Symposium 209) p 195–206

Ribozymes as trans-acting RNA inhibitory reagents

Since their discovery in 1982 ribozymes (catalytic antisense RNAs) have been widely used to suppress gene expression (for reviews see Cech 1982, Rossi 1994). The hammerhead catalytic motif has been successfully employed in many different organisms: bacteria (Sioud & Drlica 1991); plants (Steinecke et al 1992); amphibians (Cotten & Birnstiel 1989); flies (Zhou et al 1994); and mammals (Efrat et al 1994). However, ribozyme efficiency appears to be highly variable. Some ribozymes have been ineffective, while others have suppressed as much as 90% of targeted gene expression (L'Huillier et al 1992, Lo et al 1992, Bertrand et al 1994). Over the past few years, ribozymes have progressed from an intriguing subject of scientific study to therapeutic agents for the potential treatment of a fatal and devastating viral infection. Before ribozymes can be considered as generally useful therapeutic agents, there are a number of avenues that must be explored to improve the intracellular effectiveness of ribozymes. Since ribozymes are RNA molecules, the cellular rules governing RNA partitioning and stability can be applied to these molecules to make them more effective therapeutic agents. Successful therapeutic ribozyme applications will depend upon increasing our knowledge of RNA metabolism and movement, and applying this knowledge to the design of ribozymes. Understanding the factors that contribute to maximal ribozyme (antisense RNA) efficiency in mammalian cells is essential to the development of ribozyme-based therapies against infectious and genetic diseases.

The first step in this direction involves developing a detailed understanding of how ribozymes can be made to function more effectively in an intracellular environment. Since the discovery of hammerhead ribozymes, a great deal of information has been collected concerning their *in vitro* function (for reviews see Cech 1992, Symons 1992). Recently, these studies have culminated in the acquisition of three-dimensional structures for the hammerhead ribozyme (Tuschl et al 1994, Pley et al 1994, Scott et al 1995). Indeed, for short RNA sequences we are almost at the point where the *in vitro* kinetic behaviour of ribozymes can be predicted on the basis of their base-pairing sequences. Nevertheless, despite intense research, our understanding of ribozyme functioning *in vivo* lags well behind the *in vitro* knowledge, although some progress has been made towards this goal. Notably, it appears that binding of the ribozyme to its target is a crucial step *in vivo* (Sullenger & Cech 1993, Crisell et al 1993), and that this step largely determines the functional activity of the ribozyme. However, even though we recognize that a number of parameters are able to influence the intracellular functioning of ribozymes by modifying the rate of the binding of the ribozyme to its target, predicting *in vivo* efficiency remains enigmatic, and optimization relies mainly on empirical testing.

In vitro studies have resulted in a set of general rules concerning the hammerhead ribozyme reaction. First, the reaction is composed of a series of sequential steps (Hertel et al 1994, Hertel & Uhlenbeck 1995): binding of the ribozyme to the target; cleavage of the target; and release of the cleavage products. Second, simple principles govern the rate of each step: (i) the cleavage rate is relatively independent of the primary sequence of the binding arms of the ribozymes, provided the length of each arm is greater than 2 nt, and this rate is about 1 per min; and (ii) the rates of target binding and the release of each cleavage product follow the general rules of nucleic acid hybridization. Several conclusions concerning the length of the ribozyme-binding arms can be drawn. First, if this length is too short (less than 6 + 6 nt on either side of the catalytic centre for an average sequence), the rate of dissociation of the target from the ribozyme will exceed the rate of cleavage, resulting in poor turnover. Second, if the lengths of the substrate-binding arms are long (greater than the minimal length required for efficient product release), the target can be cleaved efficiently, but the ribozyme will turn over slowly due to the rate-limiting release of the cleavage products (Hertel et al 1994, Hertel & Uhlenbeck 1995). Therefore, for optimal ribozyme turnover of substrate *in vitro*, the optimal lengths of the substrate-binding arms are those that allow sufficiently stable ribozyme substrate interactions to facilitate cleavage, yet allow rapid dissociation of the cleaved products.

RNAs are constantly associated with proteins (for reviews see Citovsky & Zambryski 1993, Dreyfuss et al 1993), and it is therefore a virtual certainty that ribozymes and target RNAs will interact with a variety of proteins in an intracellular environment. Most of these proteins, such as the heterogeneous nuclear ribonucleoprotein (hnRNP) particles, bind RNA rather non-specifically, but they are able to influence RNA–RNA hybridization reactions, which led to the hypothesis that such proteins were acting as RNA chaperones by promoting the correct folding of RNAs (Dreyfuss et al 1993, Herschlag et al 1994). The kinetic effects of these proteins are twofold: they have an unwinding activity; and they possess an annealing activity that can generate up to a 1000-fold stimulation of the hybridization rate. Tsuchihashi et al (1993), Herschlag et al (1994) and we (Bertrand & Rossi 1994) have analysed the effect of such proteins, namely HIV1-encoded NCp7 and cellular hnRNP A1, on the kinetic behaviour of hammerhead ribozymes. Both studies showed that these proteins had an effect on the ribozyme reaction, and that this effect was due to the annealing and unwinding activity of these proteins. More specifically, it was shown that their annealing activity can facilitate binding of the ribozyme to its target, while their unwinding activity can promote release of both the target and the cleavage products from the ribozyme. Interestingly, we showed that each of these effects was dependent upon the length of the ribozyme arms: with short arms (6 nt), target binding was inhibited; with longer arms (7 nt), target binding was no longer inhibited but enhanced, and ribozyme turnover was facilitated; and with even longer arms (8 or more nt), ribozyme turnover was no longer activated. This effect of the ribozyme arm length is most likely a function of duplex stability. At physiological temperature and ionic strength, these proteins are able to unwind short duplexes, but not longer ones. In any case, these data show that it is possible to design the ribozyme in a way that will markedly enhance its activity in the presence of these proteins.

Ribozyme gene delivery

The effective use of ribozymes as therapeutic agents is dependent upon effective means of delivering ribozyme gene constructs to the appropriate target cells (for a review of gene therapy targeting and vectors see Morgan & Anderson 1993). Presently, retroviral vector delivery of ribozyme gene constructs looks most promising for *ex vivo* gene transfer protocols. These vectors can be engineered to harbour either Pol II or Pol III transcriptional cassettes for ribozyme expression. Highly engineered and well-defined ribozyme transcripts are desirable over longer, poorly characterized transcripts that could impede ribozyme function. Retroviral vectors can be used for the effective expression of tRNA-based Pol III transcriptional units when the transcriptional cassette is inserted within the non-transcribed region of the 5' long terminal repeat (LTR). It is also likely that other well-defined RNA polymerase II and III transcriptional units, such as those encoding the small nuclear (sn) RNAs, can also be expressed using a similar strategy, expanding the potential for gene expression in a variety of cell types.

There are some restrictions for expressing a well-defined ribozyme transcript from a retroviral vector. If a strong promoter is inserted downstream of the viral LTR, there is the possibility of promoter interference. A ribozyme construct placed under control of the viral LTR will usually be included within a long transcript. If a strong terminator is placed downstream of the ribozyme, it will terminate full length viral transcripts prematurely and in turn reduce the titre of virus produced in the packaging cell line.

Other viral vectors are being developed for gene therapy uses. Any vector that has applications for other forms of gene therapy should be suitable for the delivery of ribozyme-encoding genes. One such vector is adeno-associated virus (AAV). Although there are no published reports of AAV delivery of ribozyme-encoding genes, it has been used successfully to transduce human cells with an antisense construct that effectively blocked HIV replication in cell culture (Johnson et al 1993). Many of the constraints imposed by genome organization in retroviral vectors will not occur in AAV, but it remains to be determined how generally useful it will be for delivering genes expressed from promoters such as Pol III.

Gene therapy applications of hammerhead ribozymes

The effective use of hammerhead ribozymes in a patient setting will require effective targeting of ribozyme-containing constructs to the appropriate cells. Two major approaches are currently being explored. The first relies upon transduction of a ribozyme gene or genes into T4 lymphocytes. This can be accomplished *ex vivo*, and transduced cells can be reintroduced into the patient. Since the T cells contain HIV, non-infected cells will have to be protected with other antivirals during the transduction process to prevent *ex vivo* infectious spread of the virus. The other approach involves transduction of the precursor cell population that gives rise to T4 lymphocytes and other cell lineages that are targets for HIV infection (see below).

Ribozymes as anti-HIV1 therapeutic agents

We first successfully tested the concept of using ribozymes as a therapeutic agent several years ago by expressing intracellularly a hammerhead ribozyme targeted to a *gag* cleavage site in the HIV1 genome. Cleavage by the ribozyme resulted in up to a 40-fold reduction in viral p24 antigen production in HeLa $CD4^+$ cells challenged with HIV1 (Chang et al 1990, Sarver et al 1990). We have also successfully immunized human T cells (Zhou et al 1994) as well as primary bone marrow stem cell-derived monocytes (I. Bahner, C. Zhou, J. Rossi, D. Kohn, unpublished data 1996) against HIV1 infection with retrovirally transduced anti-*tat* and anti-*tat/rev* ribozymes. More recently, in preparation for the first human gene therapy trial using hammerhead ribozymes transduced into human haemopoietic progenitor cells, the anti-*tat* and *tat/rev* ribozymes have been transduced into the haemopoietic stem cell bearing the CD34 antigen, a progenitor of several HIV-infectible haemopoietic cells. These cells have been derived from peripheral blood obtained from HIV-infected volunteers and have been transduced with a murine-based retroviral vector that expresses the anti-*tat* and *tat/rev* ribozymes (Fig. 1). These cells can be cultured under conditions that favour their differentiation into monocytes bearing the CD4 antigen required for HIV infection. A summary of the challenge and protection from one such patient sample is also presented in Fig. 2. The double ribozyme expression system has proven to be an important strategy, since tests with HIV variants harbouring mismatched sequences to the binding arms of the ribozymes demonstrated that the efficiency of ribozyme cleavage can be significantly reduced. If one of the sites is altered, the non-affected site is still efficiently cleaved when both ribozymes are simultaneously present during the cleavage reactions. These data clearly demonstrate the importance of a multiple ribozyme approach for the treatment of HIV infection. In addition to multiple ribozymes against HIV, we are also exploring the use of a ribozyme targeted to the mRNA for a cellular receptor utilized by HIV, called the CCK5 chemokine receptor (Dragic et al 1996). Our goal is to reduce the expression of this receptor, which is apparently non-essential for human cellular function but is essential for HIV infectivity. Thus, a planned combined attack on the virus and its essential receptor should provide a powerful approach to protecting patient cells against the spread of HIV1 infection in gene therapy applications of ribozymes.

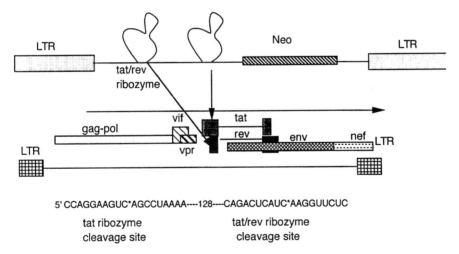

5' CCAGGAAGUC*AGCCUAAAA----128----CAGACUCAUC*AAGGUUCUC

tat ribozyme   tat/rev ribozyme
cleavage site   cleavage site

FIG. 1. Anti-*tat* and *tat/rev* ribozymes and expression vectors. The ribozymes are expressed from the murine-based retroviral vector LNL6 and their interaction with the HIV *tat* and *tat/rev* target RNAs are depicted. The upper portion of the diagram depicts the retroviral vector used to deliver the ribozymes to the human haemopoietic progenitor cells. The middle diagram depicts the regions in HIV where the ribozymes are targeted. The lower portion shows the sequences in the target to which the ribozymes bind. The asterisks indicate the sites of ribozyme cleavage. LTR, long terminal repeat.

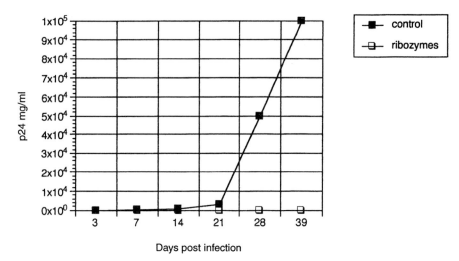

Days post infection

FIG. 2.   HIV1 challenge of HIV patient-derived monocytes expressing the anti-*tat* and *tat/rev* ribozymes. Peripheral blood-derived CD34⁺ haemopoietic progenitor cells were derived from the patient following treatments with granulocyte colony-stimulating factor. These cells were collected and transduced with the retroviral construct harbouring the anti-*tat* and *tat/rev* ribozymes. Cells were cultured on bone marrow stroma in the presence of growth factors that drive differentiation into monocytes. The resultant CD4⁺ monocytes were challenged with a monocytotropic strain of HIV1 (JR-FL) at a multiplicity of infection of 0.01. Viral production is measured as p24 antigen production (*y*-axis) versus time after viral challenge (*x*-axis). The closed squares represent the control vector lacking the ribozymes, the open squares represent the vector expressing the ribozymes.

Co-localization of ribozyme and substrate

We, and others, have made the observation that ribozyme-mediated protection of cells can be overcome with increasing multiplicity of infection. In a patient setting, this is likely to be a serious problem since there is substantial evidence suggesting that the virus is highly concentrated in the lymphoid system providing, in essence, a high multiplicity of infection to CD4⁺ cells entering that environment (Fauci 1993, Pantaleo et al 1993, Embretson et al 1993), a problem that must be dealt with in a rigorous fashion if ribozymes are to be of any real therapeutic value. Two approaches for overcoming this problem are planned. The first is the use of RNA localization signals to place the ribozyme into the same cellular subcompartment as the target RNA. The second approach involves high level expression of the ribozymes targeting the appropriate intracellular subcompartment.

From the perspective of anti-HIV ribozyme therapeutic applications, capitalizing upon the localization properties of RNAs could facilitate intracellular functioning of ribozymes by allowing them to co-localize with their target RNAs. Moreover, retroviruses have unique aspects in their life cycles that can be capitalized upon for ribozyme co-localization with the target. The life cycle includes specific events such as dimerization, packaging and reverse transcription. These events can be subverted to improve ribozyme viral RNA co-localization. A novel approach took advantage of a retroviral RNA packaging signal to facilitate intracellular co-localization of ribozyme with its target (Sullenger & Cech 1993; Fig. 2). By co-expressing murine retroviral genomic transcripts, one harbouring the *lacZ* target and the other a ribozyme to this target, an approximate 90% inhibition in viral packaging of the *lacZ*-containing viral construct was achieved as a consequence of co-localizing the ribozyme and target in the cell cytoplasm through the viral packaging signal. A mutant, non-cleaving ribozyme expressed in the same manner gave no inhibition. The expression of *lacZ* mRNA, which did not include the packaging signal, was not affected by the ribozyme,

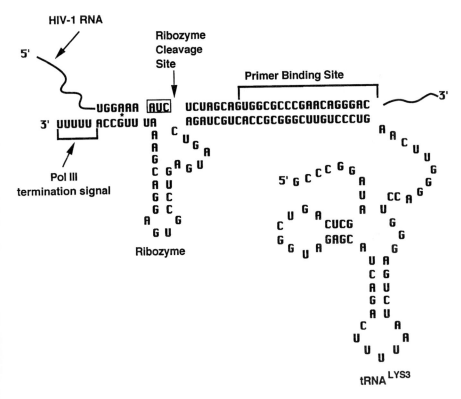

FIG. 3. Chimeric tRNA^{Lys3}–primer binding site ribozyme shown as it would pair with the primer binding site of HIV1. The asterisk marks a mismatched base pair deliberately introduced to prevent premature transcriptional termination by the RNA polymerase III enzyme which transcribes this RNA. The site of ribozyme cleavage is indicated. The chimeric tRNA–ribozyme is captured by HIV reverse transcriptase and nucleocapsid protein during viral assembly and positioned by these enzymes to the primer-binding site, where it functionally destroys the virion RNA.

dramatically highlighting the significance of the co-localization strategy for obtaining ribozyme-mediated destruction of the target RNA. This same approach could also be applied to HIV targeting, although the packaging requirements for HIV RNA appear to require more genetic information than simply the *psi* or packaging domain.

We have developed a different, but conceptually similar, strategy for co-localizing a ribozyme and target RNA (Fig. 3). Our strategy takes advantage of the fact that HIV1 utilizes $tRNA^{Lys3}$ as a site-specific primer for reverse transcription (for review see Litvak et al 1994). We have appended a ribozyme capable of pairing with and cleaving HIV1 at a site just upstream of the primer-binding site to the 3' end of $tRNA^{Lys3}$. Chimeric tRNA–ribozyme molecules can be bound by HIV1 reverse transcriptase and captured during viral assembly. We have demonstrated that: (a) the tRNA–ribozyme binds selectively to HIV1 reverse transcriptase with a binding affinity virtually identical to a synthetic $tRNA^{Lys3}$; (b) the tRNA–ribozyme is expressed as a Pol III transcript when transfected into human cells and can be overexpressed when fused to an exogenous Pol III promoter (the human U6 snRNA gene promoter); (c) the tRNA–ribozyme can localize to the cytoplasm and become packaged within the HIV virion; and (d) co-transfection of the tRNA–ribozyme gene with HIV1 proviral DNA into human embryonic cells resulted in up to a 20-fold reduction in infectious virus production relative to control constructs. Importantly, the ribozyme was only effective in inhibiting HIV replication after it was co-localized in the virion since no reductions in virally encoded proteins were observed in cells co-transfected with the ribozyme constructs and HIV proviral DNAs.

Based upon these exciting results, other strategies for co-localization combined with elevated levels of ribozyme expression are being exploited in our laboratory. It is our long-range goal to combine several strategies for ribozyme expression and targeting into a single gene delivery vehicle, thereby maximizing the potential therapeutic success of ribozymes in the treatment of HIV infection.

Acknowledgements

This work was supported by grants from the National Institutes of Health AI25959 and AI29329. I am indebted to the contributions of the following individuals whose research is summarized in this article: Shawn Westaway, Garry Larson, Laurence Cagnon, Haitang Li, John Zaia and Shirley from City of Hope; Donald Kohn, Gerhard Bauer and Ingrid Bahner from the Children's Hospital of Los Angeles; and Chen Zhou from Immusol Inc., San Diego.

References

Bertrand EL, Rossi JJ 1994 Facilitation of hammerhead ribozyme catalysis by the nucleocapsid protein of HIV-1 and the heterogeneous nuclear ribonucleoprotein A1. EMBO J 13:2904–2912
Bertrand E, Pictet R, Grange T 1994 Can hammerhead ribozymes be efficient tools to inactivate gene function? Nucleic Acids Res 22:293–300
Cech TR 1992 Ribozyme engineering. Curr Opin Struct Biol 2:605–609

Chang P, Cantin EM, Zaia JA et al 1990 Ribozyme mediated site-specific cleavage of the HIV genome. Clin Biotech 2:23–31

Citovsky V, Zambryski P 1993 Transport of nucleic acids through membrane channels: snaking through small holes. Annu Rev Microbiol 4:7167–7197

Cotten M, Birnstiel M 1989 Ribozyme mediated destruction of RNA *in vivo*. EMBO J 8:3861–3866

Crisell P, Thompson S, James W 1993 Inhibition of HIV-1 replication by ribozymes that show poor activity *in vitro*. Nucleic Acids Res 21:5251–5255

Dragic T, Litwin V, Allaway GP et al 1996 HIV-1 entry into CD4$^+$ cells is mediated by the chemokine receptor CC-CKR-5. Nature 381:667–673

Dreyfuss G, Maturis MJ, Pinol-Roma S, Burd CG 1993 hnRNP proteins and the biogenesis of mRNA. Annu Rev Biochem 62:289–321

Efrat S, Leiser M, Wu YJ et al 1994 Ribozyme-mediated attenuation of pancreatic beta-cell glucokinase expression in transgenic mice results in impaired glucose-induced insulin secretion. Proc Natl Acad Sci USA 91:2051–2055

Embretson J, Zupancic M, Ribas JL et al 1993 Massive covert infection of helper T lymphocytes and macrophages by HIV during the incubation period of AIDS. Nature 362:359–362

Fauci AS 1993 Multifactorial nature of human immunodeficiency virus disease: implications for therapy. Science 262:1011–1018

Herschlag D, Khosla M, Tsuchihashi Z, Karpel RL 1994 An RNA chaperone activity of non-specific RNA binding proteins in hammerhead ribozyme catalysis. EMBO J 15:2913–2924

Hertel KJ, Uhlenbeck OC 1995 The internal equilibrium of the hammerhead ribozyme reaction. Biochemistry 34:1744–1749

Hertel KJ, Herschlag D, Uhlenbeck OC 1994 A kinetic and thermodynamic framework for the hammerhead ribozyme reaction. Biochemistry 33:3374–3385

L'Huillier PJ, Davis SR, Bellamy R 1992 Cytoplasmic delivery of ribozymes leads to efficient reduction in alpha-lactalbumine RNA mRNA levels in C127I mouse cells. EMBO J 11:4411–4418

Litvak S, Sarih-Cottin L, Litvak M, Andreola M, Tarrago-Litvak L 1994 Priming of HIV replication by tRNA^{Lys3}: role of reverse transcriptase. Trends Biol Sci 19:114–118

Lo KMS, Biasolo MA, Dehni G, Palu G, Haseltine WA 1992 Inhibition of replication of HIV-1 by retroviral vectors expressing *tat*-antisense and anti-*tat* ribozyme RNA. Virology 190:176–183

Morgan RA, Anderson WF 1993 Human gene therapy. Annu Rev Biochem 62:191–217

Pantaleo G, Graziosi C, Demarest GF et al 1993 HIV infection is active and progressive in lymphoid tissue during the clinically latent stage of disease. Nature 362:355–358

Pley HW, Flaherty KM, McKay DB 1994 Three dimensional structure of a hammerhead ribozyme. Nature 372:68–74

Rossi JJ 1994 Controlled, intracellular expression of ribozymes: progress and problems. Trends Biotechnol 13:1–9

Sarver N, Cantin EM, Chang PS et al 1990 Ribozymes as potential anti-HIV-1 therapeutic agents. Science 247:1222–1225

Scott WG, Finch JT, Klug A 1995 The crystal structure of an all-RNA hammerhead ribozyme: a proposed mechanism for RNA catalytic cleavage. Cell 81:991–1002

Sioud M, Drlica K 1991 Prevention of human immunodeficiency virus type 1 integrase expression in *Escherichia coli* by a ribozyme. Proc Natl Acad Sci USA 88:7303–7307

Steinecke P, Herget T, Schreier PH 1992 Expression of a chimeric ribozyme gene results in endonucleotyic cleavage of target mRNA and a concommitant reduction of gene expression *in vivo*. EMBO J 11:1525–1530

Sullenger BA, Cech TR 1993 Tethering ribozymes to a retroviral package signal for destruction of viral RNA. Science 262:1566–1569

Symons RH 1992 Small catalytic RNAs. Annu Rev Biochem 61:641–671

Tsuchihashi Z, Khosla M, Herschlag D 1993 Protein enhancement of hammerhead ribozyme catalysis. Science 267:99–102

Tuschl T, Gohlke C, Jovin TM, Westhof E, Eckstein F 1994 A three-dimensional model for the hammerhead ribozyme based on fluorescence measurements. Science 266:785–789

Zhou C, Bahner IC, Larson GP, Zaia JA, Rossi JJ, Kohn DB 1994 Inhibition of HIV-1 in human T-lymphocytes by retrovirally transduced anti-*tat* and *rev* hammerhead ribozymes. Gene 149:33–39

DISCUSSION

Gewirtz: Typically, the way the CD34 mobilization is done is first to ablate the patient's marrow, because that's what increases the stem cell population, and then to give them granulocyte colony-stimulating factor (G-CSF) to release the CD34$^+$ cells into the circulation, but it seemed that this was not the protocol you were using.

Rossi: We perform the mobilization without ablation. We give a series of G-CSF injections, and at the end of this period we observe a 50–100-fold increase in circulating CD34$^+$ cells.

Gewirtz: The stem cell population is resident within the CD34$^+$ cell population, but it represents only a small fraction. This has been a problem for people trying to target retroviral vectors to the CD34$^+$ cell population. Apparently, there is poor infection efficiency of the quiescent cells (which are the stem cells). Have you looked at the infection efficiency of the stem cell population?

Rossi: The CD34$^+$ cells themselves are a heterogeneous population, and the transduction efficiency of that population varies from patient to patient. The transduction efficiency in our studies varied from 15% to as much as to 50% using this method. These are presumably all cycling cells because these vectors do not transduce to non-cycling cells. We do not know whether those cells are too far differentiated to be useful but we do know that they can differentiate into monocyte lineages in culture, and those monocyte lineages then become infectible with monocytotrophic strains (JR-FL) of the virus.

Gewirtz: That's a different branch point in haemopoetic cell differentiation.

Rossi: It may be beyond the T cell branch point, although it's not clear that we're not going to hit some cells that will differentiate into T cell lineages as well.

Gewirtz: There was a paper by Sullenger and Cech in *Science* a few years ago that discussed targeting (Sullenger & Cech 1993). They looked at the effects of a retrovirus on β-galactosidase. They found that co-localization was important and that it had to take place in the nucleus where the RNA was being processed, and not out in the cytoplasm, which is the opposite of what I understood you to say.

Rossi: Their data actually showed that RNA co-localization took place in the cytoplasm during virus assembly. They showed that the two dimerization domains in the RNA — one containing the ribozyme and one containing the target that the ribozyme was addressed to — were co-packaged in the cytoplasm into a virus particle defective in transduction (because that virus particle either had only one functional copy of RNA or it had none because the ribozyme had destroyed the β-galactosidase sequence and therefore it was transducing for β-galactosidase). We have repeated these experiments with a slightly different retroviral backbone, and we find that this kind of dimerization procedure produces the best inhibition by the ribozyme. We have also gone one step further and deleted the packaging, or *psi*, domain, which results in a loss of ribozyme activity. This activity takes place in the cytoplasm, so we can block translation by this approach.

Matteucci: How conserved is the HIV *tat/rev* sequence?

Rossi: It is highly conserved. There is little heterogeneity amongst all of the isolates we have looked at.

Matteucci: Can your ribozymes handle one or two mismatches?

Rossi: Yes. The *tat* ribozyme is not so highly conserved, but we chose it because it had a GUC site and had the correct GC content that we knew was important for interaction. The monocytotrophic strain that we challenged with has two GU mismatches in the binding arm of the *tat* ribozyme. *In vitro* we see a reduction in cleavage activity at 37 °C, but we see no reduction if we raise the temperature to 45 °C, which suggests that the changes in the virus target are structural perturbations that don't allow the ribozymes to interact. The *tat/rev* target is completely intact, so in all these challenge experiments with the monocytotrophic strains there are two mismatches to the *tat* ribozyme, and none to the *tat/rev* ribozyme, but the inhibition is the same as if it were completely matched. It can tolerate some mismatches, as long they are of a certain type and are not too close to the catalytic centre.

Inouye: The transcript harbouring the ribozyme used is 2000 bp long. Why don't you use a shorter ribozyme?

Rossi: It could be much shorter, but the 2000 bp sequence worked best in all of the early trials. We tried to make shorter, more well-defined transcripts using various promoter cassettes but none of them worked functionally as ribozymes in these transcriptional units. We have found a number of times that for an mRNA target it is best to have the ribozyme in a translatable context. The ribozyme in this case is upstream of the translated sequence.

Inouye: You have also shown that RNA-binding proteins are helpful in directing ribozymes to their target, but these proteins are in the nucleus. Do they remain associated with the ribozymes in the cytoplasm?

Rossi: Yes. They travel with the RNA from the nucleus to the cytoplasm. This is true for heterogeneous nuclear ribonucleoprotein (hnRNP) A1, which is associated with the RNA to some level in the cytoplasm, although it's not clear how long it stays associated. It may dissociate as the translational apparatus approaches the RNA.

Inouye: Does this protein help the oligonucleotides recruit RNase H?

Rossi: No. It probably helps by ironing out small kinks in target RNA, such that short helices, which might obscure or prevent base pairing, can be opened up to allow the oligonucleotides to bind. This won't work, however, for long helical regions or regions with strong ΔGs because these proteins have weak stabilizing activity.

Gait: Taira's group and others suggested linking four or five ribozymes together in a single transcript (Chen et al 1992, Ohkawa et al 1993). Have you tried similar experiments?

Rossi: No. We started with two, and we are concentrating our efforts on getting this through clinical trials.

Gait: In vitro do you get any improvements with two tandem ribozymes?

Rossi: Yes. What is probably happening is that the binding of RNA on one side opens up downstream sites, so there is more cleavage of that message than if you had the same concentration of each independent ribozyme alone. This may also be the case *in vivo*.

Monia: If the clinical trials are successful what do you anticipate the hurdles would be in the manufacturing process of these ribozymes?

Rossi: The major problem is manufacturing the vector because it has to be free of replication-competent retrovirus. This has been studied by the Chiron corporation. Ribozyme Pharmaceuticals Inc. became involved in this project and it became a way of testing a ribozyme in a clinical trial. They use a dog cell line to produce the retroviruses, which completely eliminates the potential for recombinant retroviruses. Murine cell lines are not used to produce these retroviruses because they have endogenous retroviruses to begin with, which complicates matters. In terms of scale-up, I'm not sure how valuable it would be from a financial point of view for a company because the treatment will be limited to those that can afford it. Ultimately, we would like to carry out this type of transduction using marrow from patients that have been ablated by putting transduced cells into the ablated marrow, which will increase the survival potential of the transduced cells. This is expensive and it will depend on defining the patient population in sufficient numbers that make it financially feasible.

References

Chen C-J, Banerjea AC, Harmison GG, Haglund K, Schubert M 1992 Multitarget-ribozyme directed to cleave at up to nine highly conserved HIV-1 *env* RNA regions inhibits HIV-1 replication — potential effectiveness against most presently sequenced HIV-1 isolates. Nucleic Acids Res 20:4581–4589

Ohkawa J, Yuyuma N, Takebe Y, Nishikawa S, Taira K 1993 Importance of independence in ribozyme reactions: kinetic behaviour of trimmed and of simply connected multiple ribozymes with potential activity against human immunodeficiency virus. Proc Natl Acad Sci USA 90:11302–11306

Sullenger BA, Cech TR 1993 Tethering ribozymes to a retroviral package signal for destruction of viral RNA. Science 262:1566–1569

Exogenous application of ribozymes for inhibiting gene expression

Fritz Eckstein

Max-Planck Institute of Experimental Medicine, Hermann-Rein-Strasse 3, D-37075 Göttingen, Germany

Abstract. Sequence-specific inhibition of gene expression is an attractive concept for the development of a new generation of therapeutics. Two alternatives can be envisaged for the introduction of ribozymes into cells: endogenous or exogenous delivery. In the latter, the ribozyme is prepared by chemical synthesis or transcription and delivered to the cell either unaided or with the help of liposomes. A problem with this approach is the abundance of RNases in the serum, and thus the stabilization of the ribozyme is necessary but without the impairment of catalytic efficiency. This has been achieved by several groups by 2′-modification of the pyrimidine nucleosides and the introduction of a few phosphorothioates at the termini. The selection of ribozyme-accessible sites on the target and the attachment of cholesterol and peptides to the ribozymes will be discussed. Examples of the application of these modified ribozymes in cell cultures will be presented, including the inhibition of expression of the multiple drug resistance gene, after unaided as well as liposome-aided delivery, and studies of animal models demonstrating the potential of this particular application strategy.

1997 Oligonucleotides as therapeutic agents. Wiley, Chichester (Ciba Foundation Symposium 209) p 207–217

The sequence-specific inhibition of gene expression has attracted much attention as a potential therapeutic agent. This concept has certain advantages over today's strategies as it is aimed at preventing the synthesis of a particular, harmful protein rather than using a drug that acts on it. As many disease-causing proteins, such as oncogenes, differ from the wild-type by only a few mutations, it is often difficult to develop a drug that acts specifically on the mutant protein without affecting the wild-type protein and, therefore, causing serious side-effects. The sequence-specific inhibition of translation offers the opportunity to prevent the expression of the mutant protein specifically.

The concept of this strategy was first conceived by using antisense oligodeoxynucleotides approximately 15 nucleotides in length that are complementary to a segment of the mRNA for the protein in question. The successful application of this strategy is discussed throughout this book and has also been reviewed elsewhere (Wagner 1994). A further development was the use of antisense RNAs gleaned from nature, a mechanism used by some organisms to regulate gene expression. Examples of this strategy are also dealt with throughout this book. The advent of catalytic RNA, i.e. ribozymes, offered the opportunity to tailor them for the sequence-specific inhibition of gene expression. There are several reviews describing progress in this area (Marschall et al 1994, Rossi 1995, Birikh et al 1997a, Eckstein & Lilley 1996). I will concentrate in this chapter on our work with the small, and therefore accessible by chemical synthesis, hammerhead ribozyme.

The hammerhead ribozyme

The hammerhead ribozyme (Fig. 1.) can in principle cleave any RNA that contains a $U_{16.1} H_{17}$ sequence at the cleavage site and where the annealing arms of the ribozyme are complementary to the target RNA to form helices I and III.

There are in principle two modes of delivery of ribozymes to cells: endogenous and exogenous delivery. The first consists of cloning the ribozyme gene into a vector for transfection or transduction so that the ribozyme is transcribed in the cell. The second attempts to get the preformed ribozyme into cells, often with the aid of a carrier, such as cationic liposomes. Thus, this mode of application is identical to that used with oligodeoxynucleotides, and therefore suffers from the same problems. My group follows this strategy.

Just as with oligodeoxynucleotides, the ribozymes have to be stabilized against nucleases when they are administered by exogenous delivery. We have solved this

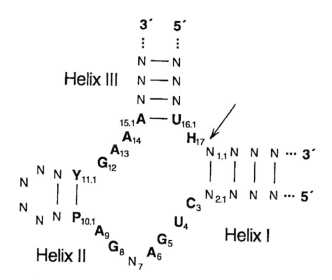

FIG. 1.   Consensus sequence of the hammerhead ribozyme. Numbering system is according to Hertel et al (1992). Conserved nucleotides are in bold. N = any nucleotide; P = A or G; Y = U or C; H = A, C or U; B = C, U or G; V = G, A or C. The arrow represents the position of cleavage.

problem by substituting all pyrimidine ribonucleotides by 2′-fluoro or 2′-amino derivatives in combination with some terminal phosphorothioates (Pieken et al 1991, Heidenreich et al 1994). Others have chosen alternative 2′-derivatives (Paolella et al 1992, Beigelman et al 1995). All these constructs are stable for several days in serum and in nuclei suspensions (Heidenreich et al 1996).

Another problem familiar to those working with oligodeoxynucleotides is the selection of suitable sites on the target RNA. As mRNAs are known to form secondary structures, it is easy to imagine that segments which contain a triplet susceptible to cleavage might not be accessible to the ribozyme. Computer fold programs are available to calculate the structure with the lowest free energy; however, it is generally agreed that the predictive power of these calculations has limitations. An experimental approach to find such sites might therefore be preferable. Lieber & Strauss (1995) have developed such a method by using a ribozyme, with randomized annealing arms, directed against a GUC triplet in human growth hormone mRNA. This approach was successful for cytoplasmic extracts from cells overproducing this RNA, and one of the selected ribozymes was further developed to inhibit gene expression in cell culture and in mice (Lieber & Kay 1996). We have chosen another route by using a completely randomized oligodeoxynucleotide of 10 nucleotides (dN_{10}) in conjunction with RNase H to identify potential ribozyme cleavage sites on human acetylcholinesterase mRNA (Birikh et al 1997b). This assay identified several sites on a transcript accessible for the dN_{10} and the RNase H. Cleavable triplets were found in the vicinity of five of these sites, and ribozymes directed against these were active in cleaving the transcript. The most active ribozyme was 250-fold more active than the best designed on the basis of the MFold computer program. Experiments to verify that these sites are also accessible *in vivo* are in progress.

Hammerhead ribozymes cleave short substrates with annealing arms not longer than five to seven nucleotides with k_{cat}/K_m efficiencies of $10–100 \times 10^6$ min^{-1} M^{-1}. Unfortunately, however, long substrates, such as transcripts, are cleaved with efficiencies that are several 100-fold lower, which is almost entirely due to the k_{react} values as the K_m values are similar to those observed with short substrates (Heidenreich et al 1994, Hendrix et al 1996) This effect is presumably due to a slow annealing step, and also to slow product release. Fortunately, certain RNA-binding proteins have been shown to improve the efficiency by 30-fold by catalysing both of these steps (Herschlag 1995). Nuclear proteins have shown similar effects (Heidenreich et al 1995). However, it is not clear what these rates are inside a cell and what rates are necessary for an efficient inhibition of gene expression. Obviously, a ribozyme could also inhibit gene expression by an antisense effect due to the complementarity of the sequences to form the ribozyme–substrate complex (Fig. 1). The importance of the catalytic effect can easily be checked by using a ribozyme where catalysis is inactivated by replacing one of the essential nucleotides in the core region.

Delivery of ribozymes to cells is still problematic. As this obstacle is identical to that for the delivery of oligodeoxynucleotides, it is not surprising that the ribozyme field follows developments in the oligodeoxynucleotide field closely. Cationic liposomes seem to be the carriers of choice at present (Leonetti et al 1993). Uptake of oligodeoxynucleotides has been shown to be facilitated by the attachment of cholesterol (Letsinger et al 1989, Krieg et al 1993). We have taken this lead to synthesize ribozymes, directed against the HIV long terminal repeat (LTR) and luciferase, which contain cholesterol that is coupled via disulfide, amide or carbamide bonds (S. Alefelder, B. Patel, F. Eckstein, unpublished observations 1996). The kinetic characteristics of these ribozymes and their power to inhibit gene expression in cell culture are at present under study. Another problem with the use of ribozymes is their slow cleavage rate of long substrates. It is believed that this is due to the slow annealing step. Recently, a peptide consisting to a large extent of lysine and arginine residues was found to accelerate the hybridization of single-stranded oligodeoxynucleotides (to form double-stranded oligodeoxynucleotides) (Corey 1995). We have attached this peptide to one of our ribozymes making use of the 2'-amino group at one of the nucleosides near the 3' terminus. We have shown previously that such an amino functionality reacts readily and specifically with aryl isothiocyanates and even better with alkyl isocyanates (Sigurdsson & Eckstein 1995, 1996). Cleavage kinetics of a long transcript with the ribozyme conjugate showed that the cleavage rate was only marginally improved, i.e. a fivefold improvement (B. Patel, F. Eckstein, unpublished observations 1996). However, these results suggest that other peptides might be coupled to a ribozyme to direct compartment localization.

Several examples show that the exogenous application of chemically modified ribozymes is a viable concept. Expression of the MDR-1 gene, which is responsible for multiple drug resistance, has been inhibited by such a ribozyme in a human cell line restoring antibiotic sensitivity (Kiehntopf et al 1994). The effect of long-chain unmodified and short-chain chemically modified HIV1 *tat-* and LTR-directed ribozymes was compared in another study (Hormes et al 1997). The explanation of this differential effect is not entirely clear at present. Proviral HIV1 DNA and the ribozymes were co-microinjected into either the nucleus or the cytoplasm of human cells. Interestingly, the long-chain ribozymes, when injected into the nucleus, inhibited viral replication, but this inhibition did not occur when they were injected into the cytoplasm. The inverse effect was seen with the short ribozymes. Even more encouraging are results obtained with *in vivo* animal models. Jarvis et al (1996) delivered a chemically modified ribozyme directed against c-*myb* with cationic liposomes to rat aortic smooth muscle cells, and this led to the inhibition of serum-induced cell proliferation. A chemically modified ribozyme directed against amelogenin, when injected into the developing molar teeth of newborn mice, temporarily prevented biomineralization (Lyngstadaas et al 1995). In another animal model Flory et al (1996) describe the decrease of stromelysin mRNA upon injection of a chemically modified ribozyme into rabbit knee joints. Stromelysin, a metalloprotease, is considered to be a mediator in arthritic disease. Surprisingly and encouragingly, in both these *in vivo* experiments the ribozymes reached their targets without the help of a carrier.

In summary, numerous examples illustrate that the concept of using preformed, chemically modified ribozymes for exogenous delivery for the inhibition of gene expression represents a viable and promising strategy, although the method requires considerably more work to be of general use.

Acknowledgements

Work in the author's laboratory was supported by the Deutsche Forschungsgemeinschaft, the Bundesministerium fur Bildung und Forschung, the German-Israeli Foundation for Scientific Research and Development and the Fonds der Chemischen Industrie.

References

Beigelman L, McSwiggen JA, Draper KG et al 1995 Chemical modification of hammerhead ribozymes: catalytic activity and nuclease resistance. J Biol Chem 270:25702–25708
Birikh KR, Heaton PA, Eckstein F 1997a Hammerhead ribozymes: structure, function and application. Eur J Biochem 3:429–437
Birikh KR, Berlin YA, Soreq H, Eckstein F 1997b Probing accessible sites for ribozymes on human acetylcholinesterase RNA. RNA 245:1–16
Corey DR 1995 48 000-fold acceleration of hybridization by chemically modified oligonucleotides. J Am Chem Soc 117:9373–9374
Eckstein F, Lilley DMJ 1996 Nucleic acids and molecular biology, vol 10: RNA Catalysis. Springer-Verlag, Berlin
Flory CM, Pavco PA, Jarvis TC et al 1996 Nuclease-resistant ribozymes decrease stromelysin mRNA levels in rabbit synovium following exogenous delivery to the knee joint. Proc Natl Acad Sci USA 93:754–758
Heidenreich O, Benseler F, Fahrenholz A, Eckstein F 1994 High activity and stability of hammerhead ribozymes containing 2'-modified pyrimidine nucleosides and phosphorothioates. J Biol Chem 269:2131–2138
Heidenreich O, Kang SH, Brown DA et al 1995 Ribozyme-mediated RNA degradation in nuclei suspension. Nucleic Acids Res 23:2223–2228
Heidenreich O, Xu X, Swiderski P, Rossi JJ, Nerenberg M 1996 Correlation of activity with stability of chemically modified ribozymes in nuclei suspension. Antisense Nucleic Acid Drug Dev 6:111–118
Hendrix C, Anne J, Joris B, Van Aerschot A, Herdewijn P 1996 Selection of hammerhead ribozymes for optimum cleavage of interleukin 6 mRNA. Biochem J 314:655–661
Herschlag D 1995 RNA chaperones and the RNA folding problem. J Biol Chem 270:20871–20874
Hertel KJ, Pardi A, Uhlenbeck OC et al 1992 Numbering system for the hammerhead. Nucleic Acids Res 20:32–52
Hormes R, Homann M, Oelze I et al 1997 The subnuclear localization and length of hammerhead ribozymes determine efficacy in human cells. Nucleic Acids Res 25:769–775
Jarvis TC, Alby LJ, Beaudry AA et al 1996 Inhibition of vascular smooth muscle cell proliferation by ribozymes that cleave c-*myb* mRNA. RNA 2:419–428
Kiehntopf M, Brach MA, Licht T et al 1994 Ribozyme-mediated cleavage of the MDR-1 transcript restores chemosensitivity in previously resistant cancer cells. EMBO J 13:4645–4652
Krieg AM, Tonkinson J, Matson S et al 1993 Modification of antisense phosphodiester oligodeoxynucleotides by 5'-cholesteryl moiety increases cellular association and improves efficacy. Proc Natl Acad Sci USA 90:1048–1052

Leonetti JP, Degols G, Clarenc JP, Mechti N, Lebleu B 1993 Cell delivery and mechanisms of action of antisense oligonucleotides. Prog Nucleic Acid Res Mol Biol 44:143–166

Letsinger R, Zhang G, Sun D, Ikeuchi T, Sarin P 1989 Cholesteryl-conjugated oligonucleotides: synthesis, properties, and activity as inhibitors of replication of human immunodeficiency virus in cell culture. Proc Natl Acad Sci USA 86:6553–6556

Lieber A, Kay MA 1996 Adenovirus-mediated expression of ribozymes in mice. J Virol 70:3153–3158

Lieber A, Strauss M 1995 Selection of efficient cleavage sites in target RNAs by using a ribozyme expression library. Mol Cell Biol 15:540–551

Lyngstadaas SP, Risnes S, Sproat BS, Thrane PS, Prydz HP 1995 A synthetic, chemically modified ribozyme eliminates amelogenin, the major translation product in developing mouse enamel *in vivo*. EMBO J 14:5224–5229

Marschall P, Thomson JB, Eckstein F 1994 Inhibition of gene expression by hammerhead ribozymes. Cell Mol Neurobiol 14:523–538

Paolella G, Sproat BS, Lamond AI 1992 Nuclease-resistant ribozymes with high catalytic activity. EMBO J 11:1913–1919

Pieken WA, Olsen DB, Benseler F, Aurup H, Eckstein F 1991 Kinetic characteristics of ribonuclease-resistant 2'-modified hammerhead ribozymes. Science 253:314–317

Rossi JJ 1995 Controlled, targeted, intracellular expression of ribozymes: progress and problems. Trends Biotechnol 13:301–306

Sigurdsson STh, Eckstein F 1995 Probing RNA tertiary structure: interhelical crosslinking of the hammerhead ribozyme. RNA 1:575–583

Sigurdsson STh, Eckstein F 1996 Site specific labelling of sugar residues in oligoribonucleotides: reactions of aliphatic isocyanates with 2'-amino groups. Nucleic Acids Res 24:3129–3133

Wagner RW 1994 Gene inhibition using antisense oligodeoxynucleotides. Nature 372:333–335

DISCUSSION

Rossi: In your random oligonucleotide RNase H experiments, where your colleague designed the ribozyme based upon those sites most accessible to RNase H, what were the arm lengths of the ribozymes? Were they comparable to the oligonucleotide or were they longer?

Eckstein: They were longer. I showed you the results for a random 10 mer, but we also did the experiment with a 15 mer and there was no difference. The ribozymes designed on the basis of these experiments had seven nucleotides in each arm, thus they were a bit longer than the random 10 mer oligonucleotide.

Rossi: The RNase H treatment is performed at 37 °C, and although the fold program is performed at a certain temperature, it doesn't take into account the binding of the oligonucleotide to various regions of the substrate. Therefore, it's difficult to predict if an AU-rich ribozyme, for example, is going to bind to a looped region at 37 °C. We have had some experience with this approach. We mixed randomized arms of ribozymes with N-labelled substrate, although this didn't work too well because of the shear numbers of ribozyme variants we had to deal with, and when we did a fold program to direct ribozymes to a particular site we found that the predicted open sites were only accessible to ribozyme pairing at 4 °C, in the case of short-arm ribozyme, because of the stability limitations of the short-armed hybrids.

Eckstein: The fold program that the Ribozyme Pharmaceuticals Inc. (RPI) group uses is more sophisticated because it takes into account, for each stretch of the RNA, its involvement in the global secondary structure and the stability of possible secondary structures not corresponding to the optimal overall structure. It also considers the ribozyme structure (Christoffersen et al 1994).

Gait: I was somewhat confused over your conjugation procedure. You mentioned attaching cholesterol or peptides through 2'-amino functional groups but you also said that you used 2'-amine groups on several positions simultaneously, so you must have changed your ribozymes somewhat.

Eckstein: The 2'-amino uridine is in the centre. However, we attach cholesterol or the peptide to ribozymes where the 2'-amino group is present only once close to the 3' terminus.

Gait: Is it reasonably efficient to make conjugates using that technique?

Eckstein: Yes, although it's not 100% efficient. We have to purify the conjugate using HPLC or FPLC.

Gait: Why is there a fivefold increased catalytic efficiency with peptide-conjugated ribozymes? Were there changes in K_m?

Eckstein: In these single turnover experiments the increased efficiency is mainly due to an increased K_{cat}. It is debatable whether the on-rate is the rate-limiting step because the on-rate actually represents multiple steps. I would say that the on-rate is not rate limiting *per se*, rather it is the conformational change between the on-rate and the subsequent steps. The factor of five in these experiments is close to the S.D. so I wouldn't like to interpret this in any sort of mechanistic way.

Cohen: Even though you retain activity with these multiple chemically modified groups, are you changing either the conformation or the interactions?

Eckstein: I assume not, although I don't have any data on this. We don't interfere with the binding to the target, and the bottom line is that the catalytic efficiency remains the same.

Letsinger: The objectives of the chemotherapy and antisense approaches are different. Is this because the antisense gene therapy approach could never result in a cure for HIV, for example?

Rossi: For HIV, which is a chronic infectious disease, the gene therapy approach could result in a cure. If the viral infection is minimized by having cells protected over a long period of time it is possible that the virus can be cleared, although this is clearly going to be difficult to prove. However, this is the goal of the gene therapy approach. The exogenous delivery approach would have to be maintained continually throughout. It also has the same potential, i.e. if the immune system can reconstitute itself in the disease it's possible to clear the virus. We should not focus entirely on therapy because there are other situations where ribozymes, and indeed antisense oligonucleotides, are useful: for example, the exogenous application of ribozymes can be used to study gene function.

Gait: To take that a step further, there should now be the possibility of comparing antisense and ribozyme approaches to the same target. Has anyone looked at this?

Iversen: In our rat liver model we looked at the efficacy of the 2'-O-allyl hammerhead ribozyme. We found that although the clearance of ribozymes was reasonably rapid, the effective molecules resided in the liver longer than traditional antisense compounds.

Eckstein: The RPI group also reported that ribozymes were more efficient than antisense oligonucleotides in their system (Jarvis et al 1996). However, if you are looking at a ribozyme site and then you direct your oligonucleotide against the same site this will bias your perception. A fairer comparison would be if one also optimized for the antisense oligonucleotide site.

Gewirtz: Is there any reason to think that one should necessarily be better than the other, precisely for the reason you just mentioned?

Eckstein: No. It is not necessarily clear that one should be better than the other.

Iversen: One consideration might be that, in terms of the specificity of the ribozyme, at least using the conservative mutation approach, this compound is absolutely inert in the rat liver. Another consideration is exposure measured by the area under the curve. It doesn't matter if the compound falls apart, as long as it isn't rapidly cleared into the urine. The race is against clearance versus breakdown.

Matteucci: In terms of whether we should use antisense oligonucleotides or ribozymes, many of us believe that oligonucleotides bind tightly to proteins in cells, such that there is a sequestration phenomenon and a small amount is free for hybridizing. Is there any evidence that ribozymes are different? Has anyone microinjected a fluorescein-tagged ribozyme to determine its cellular location and whether it is covered with protein immediately upon entering the cell?

Rossi: One experiment that may be possible to do with a ribozyme is to incorporate a signal into the non-base pairing sequences that will allow it to go back out into the cytoplasm. For instance, the U1 small nuclear (sn) RNA has a sequence motif that is involved in transporting it from the nucleus into the cytoplasm. Also, the adenovirus UA1 RNA is a small cytoplasmic RNA that has a structural motif which binds a protein that allows it to exit the nucleus.

Krieg: A related question, from someone who doesn't work on ribozymes, is what is the limiting factor in ribozyme efficacy? Is it finding the right sequence to target? Is it proteins that for some reason bind to a particular ribozyme structures?

Eckstein: But we can't even answer these questions for antisense oligonucleotides. Certainly, one of the limiting factors is that at the beginning we are not really certain which target sites are accessible. For ribozymes it has been shown that nuclear proteins improve RNA cleavage (Heidenreich et al 1995).

Gewirtz: What are the ionic requirements for magnesium ions, in terms of comparing ribozymes *in vivo* and *in vitro*?

Eckstein: Ribozymes do have a requirement for magnesium ions. The optimal concentration *in vitro* is 10 mM, whereas the free concentration of magnesium ions in cells is much lower, of the order of 200–300 μM.

Gewirtz: An oligonucleotide might work better *in vivo* because it does not have those kinds of requirements.

Eckstein: We know little about the distribution of free magnesium ions, but the ribozymes work, and we assume that this has something to do with cleavage of the RNA.

Gewirtz: But there are others who assume that this is because the flanking arms provide an antisense effect.

Eckstein: One of our control experiments involved inactivating the ribozyme by changing one of the essential nucleotides in the core. If it was working via an antisense mechanism then we would not have observed this effect.

Gewirtz: But you said that you couldn't target your antisense molecule necessarily to the same region that you're targeting the ribozymes, so those may not be the best controls for those experiments.

Rossi: It was only the catalytic core of the ribozyme that was changed and not the mechanism by which it binds.

Caruthers: One of my concerns is the low rate of catalysis for ribozymes. Is this a serious problem?

Eckstein: It doesn't seem to be, but then we know little about the catalytic efficiency within the cell. There are proteins that increase efficiency by a factor of about 30, and there may be other as yet undiscovered proteins that improve on this.

Rossi: RPI have made some modifications of the catalytic core region that result in increased catalytic efficiency, and probably also increase the rate of the cleavage step by up to 50-fold (Burgin et al 1996). They are obtaining K_{cat} values that are approaching those of a protein. In terms of site-specific cleavage activity, the ribozyme is equivalent to *Eco*RI.

Eckstein: There have been many reports where people have increased the catalytic efficiencies of ribozymes. However, when you look at the data carefully the K_m value has increased by the same amount as the increase in K_{cat}. In my opinion there is little scope for large improvements.

Agrawal: Is anything known about the cellular uptake of ribozyme RNA or linear RNA compared to DNA, and have there been any stability comparisons?

Vlassov: We investigated the efficiency of oligonucleotide uptake and found that it was more efficient for long oligonucleotides. The data suggest participation of a protein receptor in the oligonucleotide uptake.

Lebleu: There are some old studies on poly I–poly C, which is a synthetic double-stranded RNA, and these reported that uptake is relatively inefficient. Microinjected poly I–poly C is 10^4-fold more active (in terms of interferon induction) than poly I–poly C in the cell culture medium. This indicates how much can be gained by appropriate delivery.

Stein: Scavenger receptor will pick up RNA, as opposed to DNA, and this receptor although common, is not ubiquitous. Therefore, the rate at which RNA enters the cell is going to be both highly sequence dependent and cell type dependent, i.e. dependent on the density of scavenger receptor, and it is therefore not possible to generalize.

Eckstein: In terms of exogenous delivery, the RNA is highly modified and this structure may not be recognized by scavenger receptor.

Pieken: It's not clear to me how we can talk about RNA uptake without talking about stability. The rate of free RNA degradation is so fast.

Monia: I have a question about RNA oligonucleotides, as opposed to ribozymes. Under rare circumstances, we have observed a reduction in RNA levels with a RNA-based oligonucleotide, provided we stabilize it. We think that we're utilizing enzymes such as RNase III. In your controls have you seen similar reductions in RNA levels with RNA oligonucleotides?

Eckstein: We haven't, but almost certainly because we haven't looked for it.

Cohen: I have a question about the rate of the reaction. The ribozyme reaction is a transphosphorylation, but most enzyme reactions are proton transfers, which are much faster. Therefore, doesn't this suggest that there is an intrinsic rate-limiting step when comparing ribozymes to other enzymes?

Eckstein: We normally see a rate of one per minute for the chemical step. The problem is that for many enzymes people don't look at the chemical step. For instance, the rate-limiting step for restriction enzymes is not the cleavage reaction, rather it is often product release.

Inouye: You mentioned that free RNA inside the cell is not stable. However, this may not always be true. For example, we've been working on a particular mRNA in *E. coli*, in which the RNA is much less stable than other cellular mRNAs at 37 °C, and this mRNA is only induced at low temperature. This induction is controlled by mRNA stability at 37 °C, even if the gene is transcribed constitutively. The half-life of the mRNA is only 12 seconds or less. This is why the gene cannot be expressed at 37 °C. However, when we incorporate three point mutations near the Shine-Dalgarno sequence the half-life of the mRNA increases to 30 min, even at 37 °C. The mRNA has a 159 bp 5' untranslated region. It is not yet known how the stability of the mRNA in the cell is being controlled.

Eckstein: Are these point mutations at either end?

Inouye: No. They are in the middle of the gene, near the ribosome-binding site.

Toulmé: I have a question about ribozyme specificities. Ribozyme target sequences are significantly longer than those that have been used in antisense oligonucleotide studies, and the longer the antisense sequence the more likely non-specific effects will be observed because of binding to partially complementary sequences. Is there any evidence that the structure of ribozymes is so strong that it cannot unfold, or is it reasonable to think that this structure can open and bind to a region that will generate non-specific effects?

Eckstein: I am not aware of any studies on this.

References

Burgin AB, Gonzalez C, Matulic-Adamic J et al 1996 Chemically modified hammerhead ribozymes with improved catalytic rates. Biochemistry 35:14090–14097

Christoffersen RE, McSwiggen J, Konings D 1994 Application of computational technologies to ribozyme biotechnology products. Theochem J Mol Struct 117:273–284

Heidenreich O, Kang SH, Brown DA et al 1995 Ribozyme-mediated RNA degradation in nuclei suspension. Nucleic Acids Res 23:2223–2228

Jarvis TC, Alby L, Beaudry AA et al 1996 Inhibition of vascular smooth muscle cell proliferation by ribozymes that cleave c-*myb* mRNA. RNA 2:419–428

Efficient process technologies for the preparation of oligonucleotides

Wolfgang Pieken

NeXstar Pharmaceuticals Inc., 2860 Wilderness Place, Boulder, CO 80301, USA

Abstract. Efficient process technologies for the preparation of 2′-substituted nucleoside monomers, as well as for oligonucleotide preparation, are introduced. A novel method for efficient preparation of 2′-substituted uridines is presented. This method employs the 3′-hydroxyl group of 2,2′-anhydrouridine as a tether for the facile intramolecular introduction of nucleophiles to the 2′-position. It allows access to 2′-alkoxy substituents from their alcohol precursors and to substituted 2′-amino substituents, such as the novel O-substituted 2′-hydroxylaminouridines. A novel process for large-scale oligonucleotide synthesis is discussed, which allows solution phase coupling of the monomer to the growing oligonucleotide chain. This is followed by selective isolation of productive coupling product by anchoring to a resin. Release from this resin completes a coupling cycle.

1997 Oligonucleotides as therapeutic agents. Wiley, Chichester (Ciba Foundation Symposium 209) p 218–223

This Ciba Foundation Symposium documents the increased maturity of the oligonucleotide therapeutics field. Several approaches to using oligonucleotides and their modified analogues continue to develop vigorously, such as blocking gene expression by antisense oligonucleotides (Agrawal 1996) and ribozymes (Marr 1996), and blocking disease progression through oligonucleotide aptamers (Gold et al 1995). At NeXstar we have developed a variety of ligands with low nM to pM affinity for pharmaceutically relevant targets. These ligands are stabilized against nuclease degradation by 2′ modifications and typically bear a 5′ terminal polyethylene glycol substituent to increase residence time *in vivo*. Although the therapeutic applications for oligonucleotides have grown rapidly, the development of technology for preparation of these compounds at scale has not kept pace.

Many therapeutic oligonucleotides incorporate 2′ modifications for the stabilization against nuclease degradation (Pieken et al 1991, Eaton & Pieken 1995). The chemical technology for introduction of such modifications to the 2′ position of nucleosides often requires laborious differential protection of the nucleobase and the hydroxyl groups on the ribose unit. For the preparation of 2′-modified pyrimidines from their 2,2′-anhydrouridine precursor, only a limited number of analogues, such as the

2'-deoxy-2'-azidouridine (Kirschenheuter et al 1994) and 2'-deoxy-2'-fluorouridine (Codington et al 1964), are accessible. Their preparation proceeds with moderate yield under harsh conditions. The preparation of 2'-O-alkyluridine typically requires not only protection of the N^3-nitrogen but also the separation of alkylated isomers (Wagner et al 1991).

The use of a hydroxyl tether for the intramolecular region-selective introduction of a nucleophile is a well established method in the nucleophilic substitution of α-hydroxyepoxides (Roush & Adam 1985). We explored the utility of the 3'-hydroxyl group in 2,2'-anhydrouridine as a tether for the intramolecular regioselective introduction of 2'-ribose substituents (Fig. 1). A variety of linkers can be used to anchor nucleophiles to the 3'-hydroxyl group and to effect nucleophilic substitution of the 2'-position under mild conditions in high yield.

Results and discussion

We have applied this method to the preparation of 2'-O-alkylpyrimidines by nucleophilic introduction of alkoxides to the 2' position of uridine (McGee & Zhai 1996). In this case, a metal, such as magnesium, serves as the linker which anchors the nucleophile to the 3'-hydroxyl. The alkoxy substituents introduced by this method range from simple methoxy to propoxy, allyloxy and phenoxy. When the 3'-hydroxyl group is blocked by methylation, no desired 2'-alkoxyuridine derivative is isolated under identical reaction conditions. This implies that the 3'-hydroxyl group indeed complexes the metal.

This method allows access to 2'-deoxy-2'-hydroxylaminouridines, a new class of pyrimidine analogues (Sebesta et al 1996). Derivatization of the 3'-hydroxyl group of 5'-protected 2,2'-anhydrouridine with carbonyl dimidazole allows tethering of O-substituted hydroxylamines. Under mild conditions, in the presence of a base,

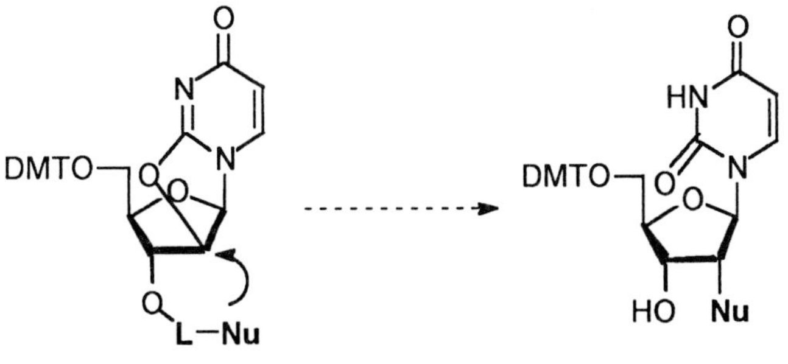

FIG. 1. Intramolecular opening of 5'-protected 2,2'-anhydrouridine by nucleophiles tethered to the 3'-hydroxyl group. A nucleophile (Nu) is attached to the 3'-hydroxyl group of 5'-protected 2,2'-anhydrouridine via a linker (L). The nucleophile attacks the 2'-position from the α-face to give, after removal of the linker group, the 2'-substituted 5'-protected uridine. DMT, 5'-dimethyoxytrityl.

the nitrogen nucleophile readily attacks the 2′ position, followed by hydrolysis of the resulting oxazolidine-one to yield the desired 2′-O-substituted hydroxylaminouridine.

Even an imidate nitrogen acts as an efficient nucleophile for 2′ substitution when tethered to the 3′-hydroxyl group (McGee et al 1996). Reaction of the 5′-dimethoxytrityl(DMT)-2,2′-anhydrouridine with neat trichloroacetonitrile results in opening of the anhydro ring to yield the cyclic trichloromethyl oxazoline intermediate, which is hydrolyzed to the 2′-deoxy-2′-aminouridine derivative under basic conditions.

Efficient monomer preparation alone does not solve the challenge of cost-efficient oligonucleotide synthesis technology at production scale. Solid-phase synthesis allows rapid preparation of small-scale amounts of oligonucleotides at the bench scale (Matteucci & Caruthers 1981). When challenged with providing larger quantities of modified oligonucleotides for drug development in a cost-effective manner, this technology shows major limitations. It is not scaleable in a predictable fashion. Much work has been devoted to solid-phase oligonucleotide scale-up. This effort resulted, over many years of work, first in the 1 mmol synthesis, and years later in 10 to 100 mmol instruments. Although these are significant achievements in scale, the adaptation of these instruments to a particular oligonucleotide compound remains laborious and costly. Furthermore, this development does not compare to the predictable and rapid scale-up of conventional synthetic organic processes.

The crude product from conventional solid-phase oligonucleotide synthesis is contaminated with highly homologous failure species, which arise from incomplete stepwise coupling. The purification of a crude oligonucleotide batch, therefore, is a highly laborious task, usually involving several high pressure liquid chromatography steps (Wincott et al 1995). This downstream processing also puts a serious limitation on scale-up and efficiency.

We developed a process for large oligonucleotide synthesis, termed product-anchored sequential synthesis (PASS), which achieves predictable scale-up typically seen in solution-phase synthetic processes, while employing the ease of solid-phase separations. The addition of the next nucleotide monomer to an oligonucleotide fragment (Fig. 2) is carried out in solution, as is typical for any organic synthetic reaction. This solution-phase reaction occurs with a minimal excess of costly monomer in high efficiency. It is amenable to straightforward scale-up and process monitoring. The monomer is derivatized (group X, Fig. 2) to allow rapid and selective anchoring on a resin, once the monomer coupling reaction is complete. This anchoring allows facile isolation of the desired reaction product from unwanted reagents and from unreacted starting material. The unreacted starting oligonucleotide fragment can be re-isolated and blended into a subsequent batch. After release of the product from the resin, one cycle of oligonucleotide synthesis is complete and a subsequent nucleotide monomer can be added.

The anchoring step provides for facile removal of the reagents and starting material. In addition, with this inherent product purification being carried out at every monomer addition cycle, the PASS process produces crude oligonucleotide products

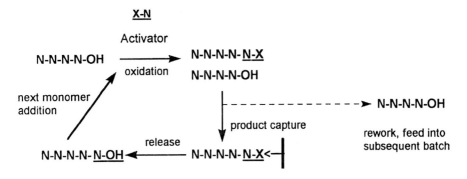

FIG. 2. Schematic of a product-anchored sequential synthesis process iteration. A monomer (N) bearing a chemical group (X) designed to react selectively and covalently with a resin is added to an oligonucleotide fragment composed of multiple N monomers. After addition of this monomer in solution, the resin is added, resulting in selective product capture. Release from the resin yields the product oligonucleotide, extended now by one monomer.

of unmatched purity. Thus, the requirement for downstream processing of oligonucleotides prepared by PASS is minimal.

Acknowledgements

I thank Bruce Eaton and Larry Gold for their critical input and continued support.

References

Agrawal S 1996 Antisense oligonucleotides: towards clinical trials. Trends Biotechnol 14: 376–382

Codington JF, Doerr IL, Fox JJ 1964 Nucleosides. XVIII. Synthesis of 2'-fluorothymidine, 2'-fluorodeoxyuridine, and other 2'-halogeno-2'-deoxy nucleosides. J Org Chem 29:558–564

Eaton BE, Pieken WA 1995 Ribonucleosides and RNA. Annu Rev Biochem 64:837–863

Gold L, Polisky B, Uhlenbeck O, Yarus M 1995 Diversity of oligonucleotide functions. Annu Rev Biochem 64:763–797

Kirschenheuter G, Zhai Y, Pieken WA 1994 An improved synthesis of 2'-azido-2'-deoxyuridine. Tetrahedron Lett 35:8517–8520

Marr J 1996 Ribozymes as therapeutic agents. Drug Discov Tod 1:94–102

Matteucci MD, Caruthers MH 1981 Synthesis of deoxyoligonucleotides on a polymer support. J Am Chem Soc 103:3185–3191

McGee DPC, Zhai Y 1996 Reaction of anhydronucleosides with magnesium alkoxides: regiospecific synthesis of 2'-O-alkylpyrimidine nucleosides. Nucleosides Nucleotides 15:1797–1803

McGee D, Settle A, Vargeese C, Zhai Y 1996 2'-Amino-2'-deoxyuridine via an intramolecular cyclization of a trichloroacetimidate. J Org Chem 61:781–785

Pieken WA, Olsen DB, Benseler F, Aurup H, Eckstein F 1991 Kinetic characterization of ribonuclease-resistant 2'-modified hammerhead ribozymes. Science 253:314–317

Roush WR, Adam MA 1985 Directed openings of 2′,3′-epoxy alcohols via reactions with isocyanates: synthesis of (+)-erythro-dihydrophingosine. J Org Chem 50:3752–3757

Sebesta DP, O'Rourke SS, Martinez RL, Pieken WA, McGee DPC 1996 2′-Deoxy-2′-alkoxylaminouridines: novel 2′-substituted uridines prepared by intramolecular nucleophilic ring opening of 2,2′-anhydrouridines. Tetrahedron 52:14385–14402

Wagner E, Oberhauser B, Holzner A et al 1991 A simple procedure for the preparation of protected 2′-O-methyl or 2′-O-ethyl ribonucleoside-3′-O-phosphoramidites. Nucleic Acids Res 19:5965–5971

Wincott F, DiRenzo A, Shaffer C et al 1995 Synthesis, deprotection, analysis and purification of RNA and ribozymes. Nucleic Acids Res 23:2677–2684

DISCUSSION

Letsinger: You did the oxidation step before capture but would there be any advantages to doing it after capture?

Pieken: It is possible to do it either way, and there may be advantages to doing the capture first. We were concerned, however, about the stability of the phosphite triester intermediate during the capture step.

Eckstein: The capacity of the maleimide cellulose column must decrease after each cycle because of the attachment of the monomer. Do you regenerate the column or take a new batch after each cycle?

Pieken: At the moment we just take a new batch, although ultimately we could regenerate it.

Fearon: What is the loading of oligonucleotide on the column, and do you observe a decrease in binding kinetics with longer oligonucleotides?

Pieken: The loading of the maleimide resin is in the mM range. We do not observe a significant decrease in binding kinetics dependent on oligonucleotide length.

Matteucci: How large are the aptamers?

Pieken: They are usually between 25 and 35 bases in length.

Matteucci: Is there a size correlation in your picomolar to nanomolar binders such that the larger the random region the better the structure?

Pieken: There is no consistent correlation between the size of the random region in the initial SELEX library and the affinity of the final ligand.

Rossi: Have you compared aptamers evolved from standard nucleosides with those from modified nucleosides, i.e. do the 2′-hydroxyl-protecting groups modify the structure significantly?

Pieken: SELEX experiments are directed at the same target but if they are performed with two libraries with different chemical composition they yield ligands of different sequences. Often, however, ribonucleoside positions in a given ligand can be substituted for 2′-modified nucleoside analogues without significant loss of affinity.

Gait: You talked about conjugation to polyethylene glycol (PEG) and how that decreases clearance. If this procedure affects other biological parameters *in vivo* is it going to be clinically useful?

Pieken: It will be useful because we see increased residence time in blood with 5'-PEG-modified ligands, which should translate into increased bioavailability.

Agrawal: Does PEG affect the secondary structure or the target affinity of the SELEX aptamers?

Pieken: The affinity for the target is usually not perturbed above the level of detection by the 5' terminal PEG modification, although it may affect hybridization to complementary sequences.

Eckstein: A critical step in obtaining these 2'-modified oligonucleotides is what is acceptable to T7 RNA polymerase or a T7 RNA polymerase mutant as substrates. What is the limiting factor and what sort of modification at the 2' position can still be tolerated for substrates?

Pieken: The 2'-O-benzyl-hydrozylamino UTP is efficiently and repeatedly incorporated by T7 RNA polymerase. It does not give a 100% product yield, rather a yield in the range of 10–20% full length transcript. This level of incorporation is surprising considering that 2'-methoxy nucleoside triphosphates and 2'-deoxy nucleoside triphosphates are not efficient substrates, although 2'-fluoro and 2'-amino nucleoside triphosphates are. My hypothesis, considering the discrimination against the proton in deoxynucleotides and the degeneracy of the system, is that this is a recognition system based on a water-mediated, hydrogen-bonding event.

Eckstein: Have you used the mutant?

Pieken: No, not yet.

Inouye: Are there any common themes in terms of secondary structures?

Pieken: There are no common themes from target to target. However, the answers to a particular SELEX experiment are generally variations of a limited number of common themes.

Fearon: You mentioned that both 2'-pyrimidine and 2'-purine modifications were necessary for *in vivo* stability but you didn't describe purine monomer synthesis. What is the progress in this area?

Pieken: The analogous approach to the intramolecular opening of 2,2'-anhydro-uridine using 2',8-anhydropurines does not give respectable yields of 2'-modified purines. Therefore, we are currently focusing on the 2'-modified pyrimidine synthesis.

Fearon: So are most of the compounds that you are currently making only 2'-modified pyrimidines?

Pieken: Yes.

Caruthers: Do these analogues affect the size of the library needed to select an inhibitor? In other words, do you have to increase or decrease the length of the random region?

Pieken: The incorporation of these analogues does not correlate with the length of the library.

Monia: How close are some of the growth factor inhibitors to being developed?

Pieken: There are animal efficacy data for many of them, and we are now choosing which ones to push forward.

In vivo production of oligodeoxyribonucleotides of specific sequences: application to antisense DNA

Masayori Inouye, Jau-Ren Mao, Tadashi Shimamoto and Sumiko Inouye

Robert Wood Johnson Medical School, Department of Biochemistry, 675 Hoes Lane, Piscataway, NJ 08854-5635, USA

Abstract. Retrons, bacterial retroelements found in Gram-negative bacteria, are integrated into the bacterial genome expressing a reverse transcriptase related to eukaryotic reverse transcriptase. The bacterial reverse transcriptases are responsible for the production of multicopy, single-stranded (ms) DNA consisting of a short single-stranded DNA that is attached to an internal guanosine residue of an RNA molecule by a $2',5'$-phosphodiester linkage. Reverse transcriptases use an RNA transcript from the retrons, not only as primer, but also as template for msDNA synthesis. By studying the structural requirement, it was found that for msDNA synthesis an internal region of msDNA can be replaced with other sequences. msDNA can thus be used as a vector for *in vivo* production of an oligodeoxyribonucleotide of a specific sequence. Artificial msDNAs containing a sequence complementary to part of the mRNA for the major outer membrane lipoprotein of *Escherichia coli* effectively inhibited lipoprotein biosynthesis upon induction of msDNA synthesis. This is the first demonstration of *in vivo* synthesis of oligodeoxyribonucleotides having antisense function. Since we have previously demonstrated that bacterial retrons are functional in eukaryotes producing msDNA in yeast and in mouse NIH/3T3 fibroblasts, the present system may also be used to produce a specific oligodeoxyribonucleotide inside the cells to regulate eukaryotic gene expression artificially. We also describe a method to produce cDNA to a specific cellular mRNA using the retron system.

1997 Oligonucleotides as therapeutic agents. Wiley, Chichester (Ciba Foundation Symposium 209) p 224–234

Bacterial reverse transcriptases are unique among all known reverse transcriptases because of their mode of priming reaction (for reviews see Inouye & Inouye 1993, 1996, Levin 1997). Bacterial reverse transcriptases are encoded by retroelements called 'retrons' found in *Myxococcus xanthus,* a minor population of *Escherichia coli,* and a number of other Gram-negative bacteria. The retron reverse transcriptases are evolutionarily related to eukaryotic reverse transcriptases and are responsible for the biosynthesis of a satellite DNA called multicopy, single-stranded (ms) DNA. msDNA consists of a short single-stranded DNA (63–163 bases) (Fig. 1) that is linked to the 2'-OH group of an internal riboguanosine residue of a short RNA molecule by a 2',5' phosphodiester linkage. The biosynthetic mechanism has been elucidated as shown in Fig. 2A. A retron is transcribed as a single transcript, and the msRNA/msDNA-coding (*msr–msd*) region of the transcript serves as a template as well as a primer for the reverse transcriptase reaction. There is an open reading frame downstream of the *msr–msd* region, which encodes reverse transcriptase. The *msr* region of the RNA transcript is unique for each retron and is recognized specifically by the reverse transcriptase from the same retron, but not from other retrons (Shimamoto et al 1995). The *msr–msd* transcript is folded by forming a stable duplex between the a2 sequence and the a1 sequence existing at the 5' end and the 3' end of the primer–template RNA, respectively. As a result, the template RNA forms a stem structure between the a1 and a2 sequence, and cDNA synthesis is initiated from the branching guanosine residue (circled in Fig. 2A) so that the cDNA synthesis using this single RNA primer–template system cannot extend any further than the branching guanosine residue, leaving the a2 sequence unused as template (see msDNA structures in Fig. 1).

The requirement of the structures in the region corresponding to DNA for msDNA synthesis has been extensively investigated (Shimada et al 1994). It was demonstrated that the upper stem region of msDNA was inessential and could be deleted to produce a truncated msDNA. This result raises an interesting question of whether the upper region of a msDNA can be replaced with other unrelated sequences. Here, we demonstrate that msDNA indeed can be used as a vector for *in vivo* production of a single-stranded DNA or an oligodeoxyribonucleotide of a specific sequence (Mao et al 1996). We further demonstrate that artificial msDNAs containing a sequence complementary to a part of the mRNA for the major outer membrane lipoprotein of *E. coli* effectively inhibited the lipoprotein biosynthesis upon induction of the msDNA synthesis (Mao et al 1996). This is the first demonstration of *in vivo* synthesis of oligodeoxyribonucleotides having antisense function. We also demonstrate that the *msr* and *msd* regions can be expressed separately and that these transcripts can form an intermolecular complex to initiate the cDNA synthesis in a cell-free system using purified reverse transcriptase (Shimamoto et al 1997). This raises the intriguing possibility that cDNA or antisense DNA to a specific cellular mRNA can be synthesized *in vivo* if the sequence directly upstream of the branching guanosine residue of the *msr* RNA is designed to be complementary to a sequence in a specific mRNA.

Production of multicopy, single-stranded DNA containing antisense DNA against the *lpp* mRNA

Retron-Ec73 is an *E. coli* retron responsible for the production of msDNA-Ec73 (Sun et al 1991). Since msDNA-Ec73 is stable and is produced at a level of approximately a few hundred copies/cell, this msDNA was used as a vector for the *in vivo* production of

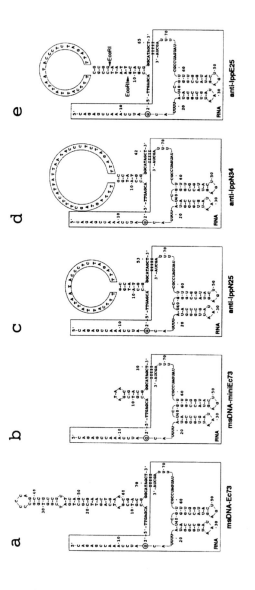

a   b   c   d   e

f

lpp mRNA : 5'ₚₚₚGCUACAUGGAGAAUUAACUCAAUCUAĜĜĜGUAUUAAUAᴀUG̲AAGCUACUAAACUGGU--3'
                                    |||||||||||||||||||||||||||||
                            3'--TTAGATCTCCCATAATTATTACTTCGATGATTT-5'

Antisense
oligonucleotides : ___a (25 bases)___
                        ___b (34 bases)___

FIG. 1.   The proposed structures of multicopy, single-stranded (ms) DNA-Ec73, its derivatives and the antisense sequences used in the msDNAs. (a) msDNA-Ec73 isolated from clinical *Escherichia coli* strain C1-23 (Sun et al 1991). (b) msDNA-miniEc73 constructed by deleting 43 bases (from C$_{15}$ to G$_{57}$) from the DNA structure of msDNA-Ec73. (c, d) msDNA-anti-*lpp*N25 and msDNA-anti-*lpp*N34, derivatives of msDNA-Ec73 containing anti-*lpp* sequences a and b (Fig. 1f) in the loop structure, respectively. (e) msDNA-anti-*lpp*E25, a derivative of msDNA-Ec73 containing anti-*lpp* sequence a (Fig. 1f) with an *Eco*RI site at the stem region. The anti-*lpp* sequences are circled. Boxes enclose msdRNA (the RNA portion of the msDNA), and the branching guanosine residues are circled. (f) The 5' end ribosomal-binding region of the *lpp* mRNA and the nucleotide sequences of antisense DNA a (25 bases) and b (34 bases). The initiation codon, AUG, of *lpp* is shown in the box, and the Shine-Dalgarno sequence is indicated by dots.

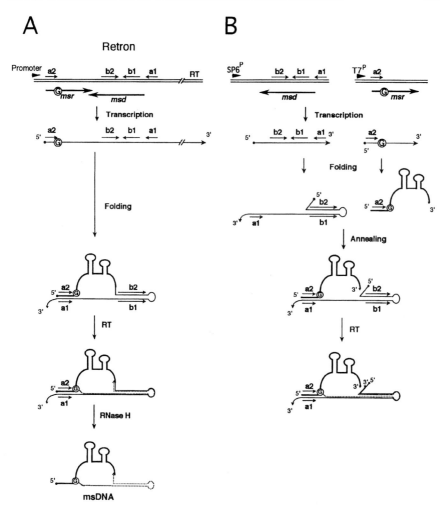

FIG. 2.   Models for cDNA synthesis by bacterial reverse transcriptase. (A) Multicopy, single-stranded (ms) DNA synthesis by a retron. (B) *msr* and *msd* RNAs are transcribed separately under different promoters (SPG and T7 promoters). cDNA synthesis is carried out by mixing two RNA transcripts. Short, thin arrows represent the inverted repeats (a1–a2 and b1–b2). Thick arrows represent the genes for msRNA *(msr)* and msDNA *(msd)*. The branching guanosine residue is circled. Long, solid lines represent the transcripts from the *msr* region required for specific recognition by reverse transcriptase (RT) (Shimamoto et al 1993) and dotted lines correspond to cDNA.

artificial oligodeoxyribonucleotides. Because we have demonstrated that the upper stem region of msDNA-Ec107 can be deleted (Shimada et al 1994), we first tested if a similar deletion is possible for msDNA-Ec73. By deleting a substantial central part of the *msd* region and using a pUC vector, we constructed a retron responsible for msDNA-miniEc73 (Fig. 1b) (Inouye & Inouye 1984). This msDNA consists of only 30 nucleotides, 43 nucleotides shorter than msDNA-Ec73 (Fig. 1a) (Mao et al 1996). The yield of msDNA-miniEc73 in the presence of 1 mM isopropyl-β-D-thiogalactopyranoside was as high as that of msDNA-Ec73, estimated to be present at 5000 copies/cell.

Because this result indicates that at least the upper part of the stem–loop structure of msDNA-Ec73 can be removed, we next attempted to add new sequences to the loop region of the msDNA-miniEc73. For this purpose, the entire stem–loop region was replaced with a NcoI site. This allowed the insertion of sequences a and b (Fig. 1f) to produce msDNA-anti-lppN25 (Fig. 1c) and msDNA-anti-lppN34 (Fig. 1d), respectively. In these msDNAs the loop sequences are complementary to the translation initiation region of the mRNA for the major outer membrane lipoprotein, the most abundant protein in E. coli. This protein was previously used as a target for antisense RNA regulation (Coleman et al 1984). Similarly, another artificial retron was constructed to produce msDNA-anti-lppE25 (Fig. 1e). This msDNA is similar to msDNA-anti-lppN25 except that it has a longer stem so that when an EcoRI site is recreated upon the formation of the msd stem structure in the msDNA, it can be digested by the EcoRI enzyme.

All the artificial retrons were constructed in the pINIII(lppP^{-5}) vector (Inouye & Inouye 1985), as in the case of msDNA-miniEc73, so that msDNA productions were inducible by isopropyl thiogalactosidase (IPTG), a lac inducer. The amounts of msDNA detected in the late logarithmic growth were somewhat different in the three constructs, possibly due to their stabilities. msDNA-anti-lppN25 was produced at the highest level and estimated to be present at 5000 copies/cell (Mao et al 1996).

Effects of antisense DNA on lpp expression

We next examined the antisense effects of the constructs on the production of the E. coli major outer membrane lipoprotein. In the presence of 1 mM, IPTG significant inhibition was detected in all the constructs (75% for anti-lppN25, 70% for anti-lppN34 and 77% for anti-lppE25) (Fig. 3). Note that some inhibitory effects were also observed, even in the absence of IPTG. This is probably due to leaky expression of msDNA in the pIN vector, as observed previously (Coleman et al 1984).

EcoRI digestion of multicopy, single-stranded DNA inside the cell

If the proposed structure in Fig. 1e is formed, then the msDNA should be digested by EcoRI. Therefore, an EcoRI site was inserted in the stem region of anti-lppE25, and when this msDNA was extracted from the cells, purified by polyacrylamide gel electrophoresis and digested by EcoRI, three single-stranded DNA fragments of expected sizes — band b, bases 12–48 (see Fig. 1a); band c, bases 49–65; and band d, bases 1–11—were observed (Fig. 4A and scheme in Fig. 4B). This indicates that the msDNA molecules form the secondary structure shown in Fig. 1e. Next, an E. coli

Plasmid none   N25   N34   E25

IPTG   -   +   -   +   -   +   -   +

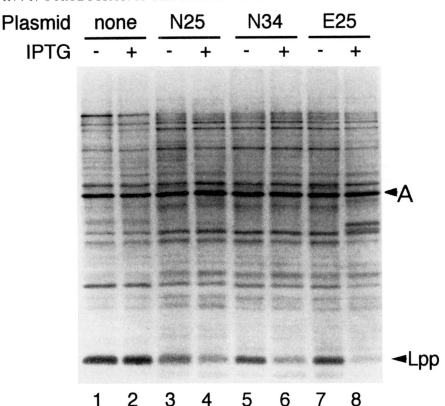

◀A

◀Lpp

1   2   3   4   5   6   7   8

FIG. 3.   Inhibition of lipoprotein (Lpp) production by antisense DNA. 1 ml cultures were labelled with 5 μCi of Trans³⁵S label (Amersham, Arlington Heights, Indiana) for 10 min at a Klett unit of 150 (Mao et al 1996). Membrane fractions were isolated by the method described previously (Coleman et al 1984) and analysed by 17.5% SDS-polyacrylamide gel electrophoresis. Lanes 1 and 2, JA221/F′*lac*I^q; lanes 3 and 4, JA221/F′*lac*I^q harbouring pINIII(*lpp*$^{p-5}$)N25; lanes 5 and 6, JA221/F′*lac*I^q harbouring pINIII(*lpp*$^{p-5}$)N34; and lanes 7 and 8, JA221/F′*lac*I^q harbouring pINIII(*lpp*$^{p-5}$)E25. Lanes 2, 4, 6 and 8 were treated with 1 mM isopropyl thiogalactoside (IPTG). The lipoprotein was quantitated using an Imaging Densitometer Model GS-670 (Bio-Rad Laboratories, Hercules, California) by comparing the density of the lipoprotein to the density of OmpA, indicated by an arrow and the letter A.

strain expressing the *Eco*RI enzyme as well as the *Eco*RI methylase was transformed with pINIII(*lpp*$^{p-5}$)anti-*lpp*E25. In this cell the chromosomal DNA is protected from *Eco*RI digestion because of the methylation at the *Eco*RI sites (Greene et al 1981). However, since the *Eco*RI site on the msDNA is formed as a result of annealing of the unmethylated single-stranded DNA synthesized by reverse transcriptase, the *Eco*RI site should still be susceptible to *Eco*RI cleavage (see the scheme in Fig. 4B). This is shown in Fig. 4A, where no intact msDNA (band a) was

A   **B**

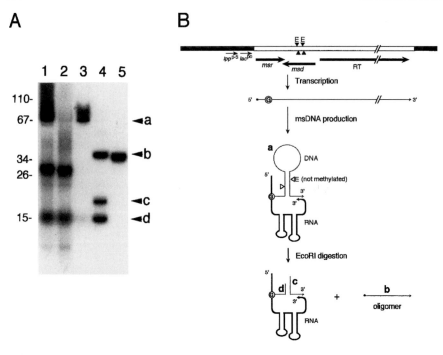

FIG. 4.   *Eco*RI digestion of multicopy, single-stranded (ms) DNA-anti-*lpp*E25. (A) Lane 1, msDNA-anti-*lpp*E25 isolated from strain MM294 (Greene et al 1981) harbouring pINIII(*lpp*$^{P-5}$)E25. The sample was treated with ribonuclease A prior to applying on a gel. Lane 2, the same fraction from strain MM294 harbouring pINIII(*lpp*$^{P-5}$)E25 and pJREcoRI that possesses an endonuclease *Eco*RI and its methylase (Greene et al 1981). pJREcoRI was constructed as previously described by Mao et al (1996). Lanes 3 and 4, gel-purified msDNA-anti-*lpp*E25 (lane 3) and its *Eco*RI digests (lane 4). All samples applied in lanes 1 to 4 were labelled at the 3′ end with [α^{32}P]ddATP and terminal deoxynucleotide transferase (TdT) (Boehringer Mannheim, Indianapolis, Indiana). Lane 5, synthetic single-stranded oligonucleotide (37 mer) used for construction of pINIII(*lpp*$^{P-5}$)anti-*lpp*E25 and labelled at the 5′ end with [γ^{32}P]ATP and T4 polynucleotide kinase (GIBCO BRL, Gaithersburg, Maryland). Sizes of the fragments a, b, c and d — which correspond to a, b, c and d in Fig. 4B — are 69 (65 deoxynucleotides and three ribonucleotides at the 5′ end plus one base), 38 (37 plus one base), 18 (17 plus one base) and 14 (11 and three ribonucleotides plus one base), respectively. Numbers on the left are single-stranded molecular weight markers. Bands seen between 34 and 26 bases, and at 15 bases, in lanes 1 and 2 are non-specific and unrelated to msDNA. (B) Genetic arrangement of pINIII(*lpp*$^{P-5}$)anti-*lpp*E25 and *Eco*RI digestion of msDNA-anti-*lpp*E25. *msr, msd* and the reverse transcriptase (RT) gene are under the control of the *lpp*$^{P-5}$ and *lac*^{po} promoters of pINIII(*lpp*$^{P-5}$). Thick arrows indicate locations and orientations of *msr, msd* and the reverse transcriptase gene. The branching guanosine is circled. Methylated and unmethylated *Eco*RI sites are indicated by filled triangles and empty triangles, respectively. An open box and solid boxes represent a retron region and pINIII vector, respectively. a, labelled full-length msDNA; b, *Eco*RI fragment that harbours the artificial msDNA loop region having antisense function; c, *Eco*RI fragment that contains the 3′ end of the msDNA; d, *Eco*RI fragment that contains the 5′ end of the msDNA.

detected, indicating that the msDNA was indeed digested by *Eco*RI. However, the expected *Eco*RI-digested products — bands b, c and d — were not detected *in vivo*. This is most likely due to their nuclease sensitivities. Nevertheless, these results indicate that msDNA may be used as a vector to produce shorter single-stranded oligodeoxyribonucleotides inside the cell.

Here we demonstrated that msDNA can be used as a vector for the *in vivo* production of a short single-stranded DNA fragment. Antisense DNA thus produced *in vivo* effectively blocked specific gene expression. Earlier we showed that msDNA can be produced in yeast (Miyata et al 1992) as well as mammalian cells (Mirochnitchenko et al 1994) by introducing a retron into these eukaryotic cells. Therefore, the bacterial retrons may also be used in eukaryotes, including plants, as a vector to produce artificial single-stranded DNAs inside the cell. Such DNAs may be designed to target not only specific mRNAs, their precursors and other functional RNAs, but also chromosomal double-stranded DNA to form triple helices (Maher 1992).

cDNA synthesis against a specific mRNA

During msDNA synthesis, cDNA synthesis is blocked at the branching guanosine residue because it physically inhibits reverse transcriptase passing through the guanosine residue. It is possible that this inhibition can be overcome if the *msr* and *msd* regions are under separate promoters so that their transcripts are produced separately (Fig. 2B). If the *msr* transcript is able to anneal to the *msd* transcript, then cDNA synthesis may be initiated at the branching guanosine residue and continued to the 5' end of the template RNA coded from *msd*. We have examined this possibility by using retron-Ec73 (Sun et al 1991) as a model system. When the primer and template RNAs corresponding to the *msr* and *msd* regions of retron-Ec73 were transcribed separately *in vitro*, these RNAs were able to function as substrates to initiate cDNA synthesis in a cell-free system using purified reverse transcriptase Ec73 (T. Shimamoto, H. Kawanishi, M. Inouye & S. Inouye, unpublished observations 1996). The 5' end sequence of the synthesized cDNA was found to be identical to that of msDNA-Ec73 and that of the cDNA synthesized *in vitro* from a single *msr–msd* RNA transcript used as a primer–template RNA (Shimamoto et al 1995). These data suggest that the *msr* transcript (primer RNA), even if separately produced from the *msd* region, is able to anneal to the *msd* transcript (template RNA) initiating the cDNA synthesis from the 2'-OH group of the internal branching guanosine residue in the *msr* transcript. This result provides an intriguing insight into a possible cellular function of msDNA in producing cDNAs to specific mRNAs. In addition, this system can be used to produce cDNA against a specific mRNA. As depicted in Fig. 5, if an mRNA contains a sequence complementary to the a1 sequence immediately upstream of the branching guanosine residue, the *msr* RNA can hybridize at that sequence and reverse transcriptase is able to prime the cDNA synthesis from the guanosine residue as shown in the figure. Since *msr* RNA can be artificially designed for a specific bacterial reverse transcriptase having a complementary sequence to a specific mRNA, one can

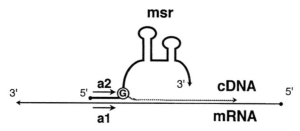

FIG. 5.   A model for *in vivo* cDNA synthesis from a mRNA containing the a1 sequence. The *msr*, *msd*, a1 and a2 regions are defined in Fig. 1. cDNA shown by a dotted line is initiated from the 2′-OH group of the branching guanosine residue in the primer RNA. The branching guanosine residue is circled.

apply this method *in vivo* in both prokaryotes and eukaryotes to block the function of the mRNA. This method may be much more effective than antisense RNA or DNA technology, since antisense DNA or cDNA is produced directly on the target mRNA in the cell.

Acknowledgement

This work was supported by a grant (GM44012) from the US Public Health Service.

References

Coleman J, Green PJ, Inouye M 1984 The use of RNAs complementary to specific mRNAs to regulate the expression of individual bacterial genes. Cell 37:429–436
Greene PJ, Gupta M, Boyer HW 1981 Sequence analysis of the DNA encoding the *Eco*RI endonuclease and methylase. J Biol Chem 256:2143–2153
Inouye S, Inouye M 1985 Up-promoter mutations in the *lpp* gene of *Escherichia coli*. Nucleic Acids Res 13:3101–3110
Inouye S, Inouye M 1993 Bacterial reverse transcriptase. In: Goff C, Skalka S (eds) Reverse transcriptase. Cold Spring Harbor Press, Cold Spring Harbor, NY, p 391–410
Inouye S, Inouye M 1996 Structure, function and evolution of bacterial reverse transcriptase. Virus Genes 11:81–94
Levin HL 1997 It's prime time for reverse transcriptase. Cell 88:5–8
Maher LJ 1992 DNA triplex-helix formation: an approach to artificial gene repressors? Bioessays 14:807–815
Mao JR, Shimamoto M, Inouye S, Inouye M 1996 Gene regulation by antisense DNA produced *in vivo*. J Biol Chem 270:19684–19687
Mirochnichenko O, Inouye S, Inouye M 1994 Production of single-stranded DNA in mammalian cells by means of a bacterial retron. J Biol Chem 269:2380–2383
Miyata A, Ohshima A, Inouye S, Inouye M 1992 *In vivo* production of a stable single-stranded cDNA in *Saccharomyces cerevisiae* by means of a bacterial retron. Proc Natl Acad Sci USA 89:5735–5739
Shimada M, Inouye S, Inouye M 1994 Requirements of the secondary structures in the primary transcripts for multicopy single-stranded DNA synthesis by reverse transcriptase from bacterial retron Ec107. J Biol Chem 269:14553–14558

Shimamoto T, Hsu MY, Inouye S, Inouye M 1993 Reverse transcriptases from bacterial retrons require specific secondary structures at the 5'-end of the template for the cDNA priming reaction. J Biol Chem 268:2684–2692

Shimamoto T, Inouye M, Inouye S 1995 Formation of the 2',5'-phosphodiester linkage in the cDNA priming reaction by bacterial reverse transcriptase in a cell-free system. J Biol Chem 270:581–588

Shimamoto T, Kawanishi H, Tsuchiya T, Inouye M, Inouye S 1997 *In vitro* synthesis of multicopy single-stranded DNA (msDNA) using separate primer and template RNAs by *Escherichia coli* reverse transcriptase. J Biol Chem, submitted

Sun J, Inouye M, Inouye S 1991 Association of retroelement with a P4-like cryptic prophage (retronphage phi R73) integrated into the selenocystyl tRNA gene of *Escherichia coli*. J Bacteriol 173:4171–4181

DISCUSSION

Gait: You displayed your antisense sequence on a loop rather than conventional antisense which is a single strand. Have you looked at the sensitivity to mutations for that antisense loop sequence?

Inouye: We haven't looked at this.

Cohen: For an effective antisense strategy a cDNA is produced against a particular target. If one could do this *in vivo* the cDNA would be distributed throughout the animal. Therefore, how would you target a particular region, as you might be able to do with an exogenous oligonucleotide?

Inouye: I have only done these experiments in *Escherichia coli*.

Krieg: But wouldn't this just involve the same gene delivery strategies as for gene therapy?

Cohen: That is exactly my point: that it has the same properties and the same problems as gene therapy.

Krieg: Presumably there is some RNA that is not reversed transcribed and converted to cDNA. Is it possible to tell whether the decreased target expression is due to that residual RNA or if it's due to an antisense effect?

Inouye: We don't have a good control for that. This is an unusual system because each transcript is making one primer template and one enzyme, and it is not known whether this enzyme turns over or is used repeatedly, although we do know that it is extremely stable.

Iversen: You showed one sequence that had the Shine-Dalgarno sequence just upstream of an initiation site. Did this sequence suppress bacterial protein expression in general?

Inouye: This was an antisense DNA to the specific gene. The region for antisense DNA is unique including the Shine-Dalgarno sequence and the initiation codon, so it only blocked this particular gene. The sequence seems to be unique enough to block only a specific gene. It is possible that one can make general antisense DNA to block bacterial cell growth using sequences against the ribosome-binding site or sequences in ribosomal RNAs.

Iversen: If you could make a bacterium that could carry a conjugating form of a plasmid this would be a fantastic antibiotic.

Inouye: We have been working on the low temperature physiology of *E. coli*. At low temperatures mRNA structures become stable, and the ribosomes translate these mRNAs with difficulty. If the temperature is lowered from 37 °C to 10 °C, for example, *E. coli* growth is blocked. During the lag phase a number of cold-shock proteins are induced. There are ribosome-specific proteins including RNA helicase and RNA chaperone proteins. All the other non-cold-shock proteins are not synthesized until cold-shock-induced ribosomal factors are synthesized, which are required for the formation of the translation initiation complex. We wondered why cold-shock proteins can still be synthesized during the lag period. We found that mRNAs for cold-shock proteins not only have the Shine-Dalgarno sequence but also the box downstream of the initiation site, which is complementary to part of 16S rRNA. We found that if RNA complementary to this region of 16S rRNA is overproduced in *E. coli* cell growth at low temperature is completely inhibited.

Krieg: Of course that couldn't be used as an antibiotic because you would have to get it into the bacteria for it to have an effect.

Caruthers: But one could imagine a vector for this being developed.

Krieg: There would still be problems in transforming the bacteria.

Iversen: But bacteria containing resistant plasmids will conjugate with bacteria that do not carry such plasmids.

Krieg: The reason that you can transform bacteria experimentally is because of strong selection for the bacteria to take up the selectable antibiotic-resistance marker, but in this case it's the opposite: the selection will be for the bacteria that are not transformed, which are usually the majority anyway. Thus, it is difficult to envision a practical application of the interesting idea to use this as an antibiotic.

Summary

Marvin H. Caruthers

Department of Chemistry and Biochemistry, University of Colorado, Boulder, CO 80309-0215, USA

Although this field can trace its origins to the initial work of Zamecnik & Stephenson (1978), it began to accelerate approximately 10 years ago. At that time enough had been accomplished to lead some of us to organize the first meetings on this subject— one by Michael Gait at Cambridge (August 1987) and the other by George Johnson of the US National Cancer Institute (a workshop at Annapolis, Maryland, US, September 1987). A major challenge at these meetings was simply to define the field. For example, could mRNA be transcribed in reverse with this RNA being capable of inhibiting mRNA translation? Alternatively, perhaps synthetic DNA could be used to form inhibitory complexes with DNA (triplexes) or mRNA (duplexes). At that time, although RNase H was known, it was not considered seriously as an integral part of the methodology that would lead to the major pathway for executing the antisense concept. Instead, inhibition of gene expression through some blocking mechanism was considered the most attractive approach. Certainly, the *in vivo* expression of an antisense oligomer appeared to be the best possible method for preparing an antisense compound, as nothing but pessimism prevailed on the overall cost and other challenges (purity and scale-up of chemical methodologies) associated with the use of chemically synthesized antisense oligonucleotides. Also, in 1987 little was known about uptake of small synthetic DNAs by cells and it was considered a strong possibility that most antisense analogues would be toxic. Hence, the prevailing opinion was simply to use antisense sequences expressed *in vivo* as these natural oligomers would be non-toxic, protected against nucleases in some manner and already present inside cells. As can be seen from the presentations at this symposium, the field has moved significantly and dramatically in a positive direction during the past 10 years. Much of the pessimism has been replaced with excellent science that addresses many of these issues. Moreover, several clinical trials have been initiated or completed and the results are positive.

At the beginning of the symposium, when the structure and synthesis of antisense DNA was the topic of discussion, Mark Matteucci perhaps best summarized our current thinking with the comment 'phosphorothioates are still king'. Simply put, the comment emphasizes that this analogue remains an essential component of every successful antisense experiment *in vivo*. This is because phosphorothioates stimulate RNase H to recognize the antisense–mRNA complex and degrade the mRNA component, which is still the only successful *in vivo* antisense DNA approach. However, as also became evident from this presentation, we have been busy synthesizing a large number of neutral and anionic DNA derivatives in the attempt to either broaden the base of potentially useful analogues or overcome the presumed limitations of the phosphorothioate backbone. An intriguing new approach to thinking about antisense analogues was also introduced, which I am sure will dominate much of our efforts over the next few years. This concept is to pre-order the unbound antisense oligonucleotide to resemble the conformation it will possess when bound to its RNA target. Proper constraint could lead to an entropic advantage, enhanced affinity and perhaps improved biological activity. Accomplishing these objectives through proper conformational restrictions will be the synthetic challenge. They can perhaps be achieved through either proper covalent modification of the DNA backbone or by non-covalent interactions, such as stacking forces between properly designed heterocyclic bases. As an example of this concept, we were introduced to antisense derivatives having 5-propynyl pyrimidines and phenoxazine rings. Because these analogues formed stable complexes with natural RNA at low concentrations and reduced length (less than 10 nucleotides), they are attractive for research in this field. Undoubtedly, they will cost less to manufacture, be active at lower concentrations and probably exhibit different, and perhaps favourable, pharmacological profiles. One criticism of this type of research is that oligonucleotides must be at least 18 mer in length in order to bind uniquely to a given gene sequence. Of course, given the complexity of the human genome, one can calculate that this is a reasonable assumption. However, in any given cell, the target is mRNA that is transcribed from only a small fraction of the total DNA. Additionally, mRNAs are present in tertiary structures where most of the sequence elements are not accessible as targets for an antisense oligonucleotide. This can be readily seen from the presentations of Stanley Crooke and Brett Monia, where systematic screens of target mRNAs with antisense oligomers reveal that most of the molecule is non-accessible toward cleavage by RNase H. As a consequence of such studies, it is clear that short antisense oligomers, perhaps less than 10 mer in length, may well uniquely interact with a target mRNA. Studies such as these are encouraging and strongly suggest that there are many possible analogues that should still be tested, as some could prove useful in the antisense field.

Perhaps a few additional comments on target selection are appropriate. Once it became clear that mRNA would be the major target for antisense oligonucleotides, the next challenge was to identify regions of this molecule that would be particularly susceptible to hybridization and subsequent cleavage by RNase H. Initially, most researchers focused on translation initiation regions with less attention being directed toward splice junctions and the 3′ untranslated region. However, more recent work, as outlined by Brett Monia, has shown that the total mRNA must be scanned with appropriate oligonucleotides as some of the most susceptible targets may not reside at any of these sites. This of course is a difficult task for the average worker, as only a few of us have access to synthetic oligonucleotides on the scale required for this screen. It was therefore of considerable interest for many of us to hear about the recent developments in Ed Southern's laboratory, where they have learned how to use arrays of oligonucleotides complementary to all regions of an mRNA in order to probe simultaneously which of these oligomers hybridize optimally to the target. Also encouraging was their observation of a good correspondence between the ability of an oligonucleotide to bind to its target and its activity as an antisense agent both in tests *in vivo* and *in vitro*. When coupled with the large amount of genome sequences that are now becoming available on an increasing number of genes, one can imagine that the identification of potential targets for studying the function of undefined genes via an antisense approach will soon become routine.

So far in this discussion I have attempted to focus on potential breakthrough discoveries that should broaden our ability to use the antisense concept generally to probe gene function. Surprisingly, the development of methods for the large-scale production of oligonucleotides, as presented to us by Karen Fearon and Wolfgang Pieken, may also fit into this category. Now that kilogram quantities of DNA can be prepared cheaply and in high purity, we can design and implement whole new classes of experiments involving animals and cell culture, where the amount of oligonucleotide ceases to be the limiting step. As has been the case with other technological accomplishments in a broader context, such as DNA sequencing and synthesis, this development should lead to rapid advances in the antisense field.

Unexpectedly, the most challenging research area has been studying antisense effects with cells in culture. Early work almost without exception showed that DNA was delivered transiently to cytoplasmic vesicles followed by release to the surrounding media without delivery in detectable amounts to the cytoplasm or nucleus. Over time this problem was solved by using cationic liposome–DNA complexes for delivery. Richard Wagner more recently synthesized cationic liposome derivatives that are far less toxic and generally more useful for delivery of DNA to most cell lines in culture. Others at this symposium, such as Bernard Lebleu, told us about a clever approach for delivering DNA to cells via natural fusogenic peptides, such as the one found as part of the coat protein of the influenza virus. An alternative approach that appears to be promising is to deliver antisense DNA as a prodrug. Promising results in this area were the use of a β-mercaptylacyl phosphate ester, as described by Brett Monia, and phosphate triesters presented by Sudhir Agrawal. Both sets of data suggest that antisense oligonucleotides delivered as these prodrugs were 20–30% orally available, which would represent a significant pharmacological advance in the development of therapeutic antisense compounds. In contrast to this type of delivery, Cy Stein has learned that DNA can enter cells through adsorptive endocytosis and pinocytosis—primarily by interacting with heparin-binding proteins such as fibroblast growth factors, vascular endothelial growth factor, CD4 and Mac-1, which is a cell surface integrin found on human polymorphonuclear leukocytes and macrophages. At this time, it is not clear whether any of these cell surface protein–DNA interactions can be used to transfer DNA to the proper cell compartment. The complexity of this problem was further emphasized by at least two laboratories where G_4-containing oligomers were shown to generate antisense-type effects simply by interacting with cell surface proteins in a non-antisense, but sequence-specific, manner (Cy Stein, K-BALB fibroblasts; and Eric Wickstrom, Burkitt's lymphoma). Clearly, much remains to be learned in this area before we can reliably and predictably transfer DNA into cells in culture.

As this field evolves, the most critical questions relate to how antisense oligonucleotides are performing in the clinic. The results are exciting. As we learned from Stanley Crooke, Brett Monia and Alan Gewirtz, patients are responding well in every respect. First, of course, patients are being treated successfully. For example, a Phase I/II study of intravitreally administered oligonucleotide in AIDS patients with refractory cytomegalovirus (CMV) retinitis gave excellent results. The test oligonucleotide (ISIS 2922) produced a dose-dependent inhibition of progression of CMV retinitis in patients who had failed all other CMV therapy, had median CD4 counts of 4 and had a median duration of CMV retinitis of 11 months. Perhaps of even more significance, the Isis (Stanley Crooke, Brett Monia) and Hybridon (Sudhir Agrawal) groups discussed the pharmacological and toxicological properties of these compounds. Both groups report excellent bioavailability, peripheral tissue distribution and clearance that support once-a-day or every-other-day dosing. They have also studied the toxic liabilities of these phosphorothioates. These include effects on clotting, complement activation and possibly cytokine release. The dose-limiting toxicities relating to these effects indicate that the therapeutic index will be satisfactory. Recent research from these laboratories with second-generation oligonucleotides (a central core of phosphorothioates flanked by either 2'-methyl or 2'-methoxyethyl phosphate nucleotides) administered to animals are also encouraging, as the biodistribution was markedly different and pharmacologically encouraging. Perhaps most importantly, these laboratories reported potent antisense activities in animals in which all the data are consistent with an antisense mechanism. We owe these pioneers a vote of thanks for showing us how to set up a proper antisense experiment.

Before closing, I would like to spend a few moments discussing ribozymes and their potential as therapeutic drugs. When we first began meetings such as this 10 years ago, no-one even mentioned the possibility of using RNA as ribozymes for this purpose. The problems seemed formidable. These included the nuclease lability of RNA, target selection with sequence-defined catalytic RNA and, of course, the challenge of large-scale chemical synthesis of RNA. At this symposium, we had two excellent presentations that allay many of these criticisms. Fritz Eckstein showed us how to stabilize catalytically active ribozymes by introducing 2'-amino, 2'-fluoro and phosphorothioate linkages at selected, non-critical ribozyme sites. When microinjected into HIV-transformed cells, these ribozymes were active toward disabling virus production and were stable against nuclease digestion. In a related study, John Rossi explained how ribozyme genes designed to inactivate the AIDS virus could be delivered via retrovirus vectors to haemopoietic progenitor cells. These cells were then protected by a challenge with HIV. This experiment reminds one of our first proposals some years ago where we were planning to block mRNA by preparing the complementary RNA *in vivo*. One can only look forward with eager anticipation to further antisense developments in the ribozyme field.

In closing, I should like to thank the speakers, who have led the discussions and who are truly experts in their fields. The discussion by the entire membership has been extremely stimulating, critical and valuable to all of us as we return home and plan our next study. Perhaps an unusual aspect of this symposium was that we have known one another both personally and professionally for many years. In retrospect, this has contributed positively toward the symposium, as we have been able to have critical untempered discussions on every subject of the sort one usually reserves for only rare occasions and with a few friends. I suspect we have learned more about our field from one another during the last three days than usually occurs from attending many meetings or reading all the latest manuscripts.

Reference

Zamecnik PC, Stevenson ML 1978 Inhibition of rous sarcoma virus replication and cell transformation by a specific oligodeoxynucleotide. Proc Natl Acad Sci USA 75:280–284

Index of contributors

Non-participating co-authors are indicated by asterisks. Entries in bold type indicate papers; other entries refer to discussion contributions.

Indexes compiled by Liza Weinkove

Subject index